The Forrest Mims Engineer's Notebook

Forrest M. Mims, III

Technology Publishing
An Imprint of Elsevier
Eagle Rock, Virginia

Permissions may be sought directly from Elsevier's Science and Technology
Rights Department in Oxford, UK. Phone (44) 1865 843830, Fax: (44) 1865
853333, e-mail: permissions@elsevier.co.uk. You may also complete your
request on-line via the Elsevier homepage: http://www.elsevier.com by selecting
"Customer Support" and then "Obtaining Permissions".

ISBN - 13: 978-1-878707-03-1

ISBN - 10: 1-878707-03-5

Library of Congress Catalog Card Number:
    91-77457

NOTE: No circuit in this book is intended for use in any life-support
system, nor in any application in which life or property may be subject to
injury. It is your responsibility to determine if use, manufacture, or sale of
any device incorporating one or more circuits in this book infringes on any
patents, copyrights, or other rights.

Transferred to Digital Printing 2009.

Technology Publishing
An Imprint of Elsevier
Eagle Rock, Virginia

CONTENTS

FOREWORD

It's a real shame that the millions who have read Forrest's articles, columns, and books over the years have never had the chance to get to know him personally. I've been lucky enough to be able to call him my friend for several years. This book is special to me because the first edition of this book was the reason why I met and got to know Forrest.

I first met Forrest in 1979. I can't recall the exact date, but it was a fearsomely hot mid-summer day in Fort Worth, Texas. I was working at Radio Shack's national headquarters in their technical publications department. My boss, Dave Gunzel, had spearheaded an effort for Forrest to generate a book of IC applications circuits that were similar to Forrest's actual working laboratory notebooks. Forrest was carefully preparing each page by hand on transparent Mylar sheets using a fine-tip pen. I monitored his progress eagerly, and one day Dave told me that Forrest Mims would be arriving the following week with the last of his Mylar originals.

Wow! I was going to really meet Forrest Mims! I hadn't seen a photo of Forrest before, nor had Dave told me much about how he looked or acted. (In retrospect, I now realize that was deliberate on Dave's part—he wanted me to "discover" Forrest on my own.) I had my own mental picture of Forrest, though. Obviously, a serious fellow. Anyone who came up with all those electronic circuits couldn't have much time for laughter. Probably sharply focused and not interested in anything other than electronics. An older gentleman, certainly, with a white beard and a fondness for jackets with elbow patches. A pipe and slight Germanic accent were also likely. He would probably think I was really stupid and not have a lot of patience with me.

The Big Day quickly arrived. Forrest was due in that afternoon. I had carefully rehearsed my welcoming speech: "Hello, Mr. Mims. It is certainly good to see you. Would you like an ashtray for your pipe?"

I was alone in the technical publications office that afternoon when someone I didn't recognize stuck his head into the office doorway. He was wearing normal business attire, smiled easily, spoke with a slight Texas accent, and was looking for Dave Gunzel. Oh brother, I thought, another new employee who's lost in Tandy Center. Doesn't this guy know that Forrest Mims is going to show up this afternoon and I don't have any time to waste on him??? I mumbled something about Dave being gone for a few minutes and that we were expecting a visitor later that afternoon.

The stranger seemed apologetic. He didn't want to waste any of my time or Dave's if we were expecting someone important, he said; he just needed to tend to a couple of matters quickly and wouldn't bother us any further. He approached my desk and extended his hand toward me.

"Hi," he said, "I'm Forrest Mims; you must be Harry."

I don't recall my reply, but I think it was the unmistakable sound of self-mortification.

Forrest had work to do and wondered if I could help. He needed to spray the Mylar sheets with a protective coating before turning them over to us for printing. We commandeered a vacant area of the then-new Tandy Center, spread out the Mylar sheets, and spent the next couple of hours emptying aerosol cans of clear lacquer while discussing the state of the universe.

That afternoon, I discovered what a remarkably unpretentious guy Forrest is. Here was one guy who had earned the right to a massive ego, yet he was straightforward, down to Earth, and almost skeptical of his achievements. Our conversation ranged from electronics to lasers to politics to Texas history to computers to religion to. . . . well, you name it. It was incredible how many subjects Forrest was interested in, and how insatiable his curiosity was about everything in the natural world. By the end of that afternoon, I felt as if I had known Forrest for years.

A lot has happened since then. I eventually left Texas to become a book editor in New York and, a few years after that, moved to California where I became a founding partner in HighText Publications. Throughout, Forrest remained a valued friend and trusted confidant. The intelligence and insight that are apparent in his circuits extends to many other areas, and he has a wit and sense of irony that are delicious. While we don't get to spend much time physically in each other's company, it's a rare week when we don't have at least two or three lengthy telephone conversations. As technology has advanced, so have our modes of interaction; we often exchange a couple of faxes per day on various subjects.

We had no idea that the book we worked on back on that hot summer afternoon in 1979 would go on to sell over 750,000 copies in its various editions. Some of the pages we worked on back in 1979 appear in this book, a testimony to the enduring quality and relevance of Forrest's work. For readers such as yourself, this book will be a valuable reference to contemporary, real-world IC applications. For me, it brings back a lot of good memories. And, no, Forrest doesn't smoke a pipe, doesn't wear jackets with elbow patches, and doesn't have a beard.

Harry L. Helms

INTRODUCTION

Since my student days at Texas A&M University I have kept a series of laboratory notebooks. In these notebooks I record details about experiments, measurements, and new ideas. Also included are many electronic circuit diagrams. Dave Gunzel, formerly the director of technical publications at Radio Shack, took an interest in my notebooks in the mid-1970s and suggested that Radio Shack might someday want to publish a book of electronic circuits based on their hand-drawn format. Several years later, Radio Shack assigned me to produce Engineer's Notebook, a 128-page book of electronic circuits. The book soon became a Radio Shack bestseller. As new integrated circuits were added to Radio Shack's product line and others were dropped, I revised the book as necessary. Later, Radio Shack authorized me to do an edition of the book for McGraw-Hill.

This revised edition for LLH Technology Publishing represents the best and most interesting circuits from all previous editions.

The integrated circuits described in this book remain among the most popular ever introduced. Most of them are readily available from Radio Shack, electronic parts suppliers, and mail order dealers. Magazines such as *Radio-Electronics* carry ads from mail-order IC dealers. A few of the chips are specialized and finding sources for them may be more difficult. Four of the devices—the CEX-4000, S50240, PCIM-161, and SAD-1024—may be available only from dealers in surplus and discontinued ICs. However, the overwhelming majority of chips described in this book are readily available from many different sources. In fact, prices for some of the more common devices have fallen substantially since the first edition of this book. Some are available today for pennies!

Most of the part numbers given for the integrated circuits in this book are generic, and various manufacturers may add additional letters or numbers or even use a completely different number. For example, the 4011 is a quad of CMOS NAND gates. An "A" suffix (4011A) means this chip can operate from a 3- to 12-volt supply. A "B" suffix (4011B) means the chip can operate from 3- to 18-volt supply. The high-voltage version of the chip is by far the most common. National Semiconductor adds a CD prefix to its versions of the 4011B (CD4011B), while Motorola adds an MC1 prefix (MC140111B). Nevertheless, both chips are functionally identical.

For additional information about chip identification and specifications, see the data books published by the various integrated circuit manufacturers. These books are available directly from manufacturers of integrated circuits and from industrial supply companies that represent integrated circuit manufacturers. They are also available from some mail-order electronics parts dealers.

Forrest M. Mims III

ABOUT THE AUTHOR

Forrest Mims has been an electronics hobbyist since building a one-tube radio kit at the age of 11. Following graduation from Texas A&M University in 1966 and service as a photointelligence officer in Vietnam, he worked for three years with high-powered lasers, solid-state instrumentation, and trained monkeys with the Air Force Weapons Laboratory in New Mexico. Since becoming a full-time writer in 1970, he's written several hundred magazine articles and scholarly papers. His articles and columns have appeared in virtually every significant electronics magazine, including *Popular Electronics, Radio-Electronics,* and *Modern Electronics.* His articles on other scientific topics have appeared in a wide range of other publications, including *National Geographic World, Science Digest, Highlights for Children,* and *Scientific American.* His editorial exploits have included an assignment from the *National Enquirer* to evaluate the feasibility of eavesdropping on Howard Hughes by laser (it was possible, but Forrest declined to take part) and getting dropped by *Scientific American* as their "The Amateur Scientist" columnist because he admitted to the magazine's editors that he was a born-again Christian. His book sales total in the millions, and he is likely the most widely-read electronics writer in the world.

Forrest is currently busy as the founding editor of *Science PROBE!,* a new magazine aimed at amateur scientists. In this role, Forrest is creating the sort of magazine that he wishes had been available in his youth while acquiring a new understanding of the frustrations of being an editor. He still keeps up a hectic pace of electronics and science experimentation and writing.

Forrest and his wife Minnie have three children, and they currently live in the Texas countryside near San Antonio. They are active in church activities, and Forrest is a Baptist deacon. He has his office and electronics lab in an old restored farmhouse adjacent to his home.

PARTS SOURCES

The chips and related components (resistors, capacitors, etc.) used in this book are available from a variety of sources, including electronics stores (such as Dick Smith Electronics in Australia and David Reid stores in New Zealand), advertisers in electronics magazines, and industrial electronics suppliers. Some chips—such as the SN76477N, SN76488N, and SAD-1024A—are a bit "rarer" and you may have to look for them at companies specializing in surplus and discontinued devices.

Manufacturers of integrated circuits publish "data sheets" giving the bare-bones specifications for a chip and "applications notes" that give additional information, including circuit schematics using the chip. These can be obtained by contacting the national headquarters of the chip manufacturer or their nearest sales office.

The manufacturer of an integrated circuit is identified by a prefix in front of the actual part number. For example, "LM741" and "MC741" would both indicate the device was the common 741 operational amplifier found on pages 93 to 96 of this book. However, the "LM" would indicate the device was manufactured by National Semiconductor while the "MC" would denote a device manufactured by Motorola. Here are some common prefixes and manufacturers:

AD	Analog Devices		M	Mitsubishi
Am	Advanced Micro Devices		MB	Fujitsu
Bx	Sony		MC	Motorola
CA	RCA (now Harris)		MM	Motorola
CD	RCA (now Harris)		NE	Signetics
Cx	Sony		PM	Precision Monolithics
DM	National Semiconductor		T	Toshiba
F	Fairchild (now National Semiconductor)		TL	Texas Instruments
			TMS	Texas Instruments
FSS	Ferranti		XR	Exar
HA	Harris		µPB	NEC
HA	Hitachi			
HD	Hitachi			
HG	Hitachi			
HI	Harris			
IR	Sharp			
KA	Samsung			
LF	National Semiconductor			
LM	National Semiconductor			
LT	Linear Technology			

REVIEWING THE BASICS

INTRODUCTION

"Can I use a 0.22 µF capacitor instead of a 0.01 µF unit?"

"Is it okay to substitute a 12,000 ohm resistor for a 10,000 ohm unit?"

This section will tackle these common questions and many others. Master them, and you will be well prepared to tackle the circuits in this book.

RESISTORS

Resistors limit the flow of electrical current. A resistor has a resistance (R) of 1 ohm if a current (I) of 1 ampere flows through it when a potential difference (E) of 1 volt is placed across it. In other words:

$$R = \frac{E}{I} \quad (or) \quad I = \frac{E}{R} \quad (or) \quad E = IR$$

These handy formulas form Ohm's law. Memorize them. You will use them often.

Resistors are identified by a color code:

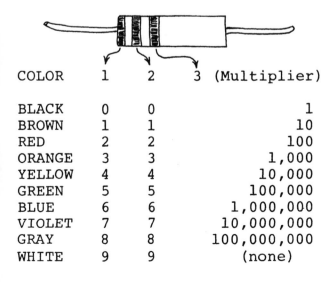

COLOR	1	2	3 (Multiplier)
BLACK	0	0	1
BROWN	1	1	10
RED	2	2	100
ORANGE	3	3	1,000
YELLOW	4	4	10,000
GREEN	5	5	100,000
BLUE	6	6	1,000,000
VIOLET	7	7	10,000,000
GRAY	8	8	100,000,000
WHITE	9	9	(none)

A fourth color band may be present. It specifies the tolerance of the resistor. Gold is ± 5% and silver is ± 10%. No fourth band means ± 20%.

Since no resistor has a perfect tolerance, it's often okay to substitute resistors. For example, it's almost always okay to use a 1.8K resistor in place of a 2K unit. Just try to stay within 10-20% of the specified value.

What does K mean? It's short for 1,000. 20K means 20 x 1,000 or 20,000 ohms. M is short for megohm or 1,000,000 ohms. Therefore a 2.2M resistor has a resistance of 2,200,000 ohms.

Resistors which resist lots of current must be able to dissipate the heat that's produced. Always use resistors with the specified power rating. No power rating specified? Then it's usually okay to use 1/4 or 1/2 watt units.

Almost every electronic circuit uses resistors. Here are three of the most important applications for resistors:

1. Limit current to LEDs, transistors, speakers, etc.

2. Voltage division. For instance:

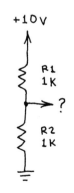

The voltage at ? is I x R2. I means the current through R1 and R2. So I = 10/(R1 + R2) or 0.005 amperes. Therefore, ? = (0.005) x (1000) or 5 volts.

Note that the total resistance of R1 and R2 is simply R1 + R2. This rule provides a handy trick for making custom resistances.

Voltage dividers are used to bias transistors:

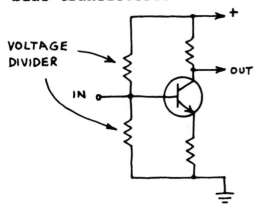

VOLTAGE DIVIDER

IN

OUT

+

They're also a convenient source of variable voltage:

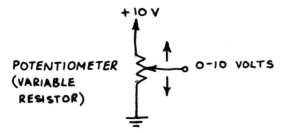

POTENTIOMETER (VARIABLE RESISTOR)

+10 V

0-10 VOLTS

And they're useful in voltage sensing circuits. See the comparator circuits in this notebook.

3. They control the charging time of capacitors. Read on...

CAPACITORS

Capacitors store electrical energy and block the flow of direct current while passing alternating current. Capacitance is specified in farads. One farad represents a huge capacitance so most capacitors have values of small fractions of a farad:

1 microfarad (μF)= 10^{-6} farad
1 picofarad (pF)= 10^{-12} farad
or
1 μF = 1,000,000 pF

The value of a capacitor is usually printed on the component. The μF and pF designations may <u>not</u> be present. Small ones marked 1-1000 are rated in pF; larger ones

marked .001-1000 are rated in μF.

Electrolytic capacitors provide high capacity in a small space. Their leads are polarized and must be connected into a circuit in the proper direction.

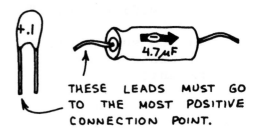

+.1

4.7 μF

THESE LEADS MUST GO TO THE MOST POSITIVE CONNECTION POINT.

Capacitors have a voltage rating. It's usually printed under the capacity marking. The voltage rating <u>must</u> be higher than the highest expected voltage (usually the power supply voltage).

Caution: A capacitor can store a charge for a considerable time after power is removed. This charge can be dangerous! A large electrolytic capacitor charged to only 5 or 10 volts can melt the tip of a screwdriver placed across its leads! High voltage capacitors can store a lethal charge! Discharge a capacitor by carefully placing a resistor (1K or more; use Ohm's law) across its leads. Use only one hand to prevent touching both leads of the capacitor.

Important capacitor applications:

1. Remove power supply spikes. (Place 0.01-0.1 μF across power supply pins of digital ICs. Stops false triggering.)

2. Smooth rectified AC voltage into steady DC voltage. (Place 100-10,000 μF across rectifier output.)

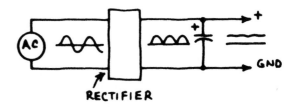

RECTIFIER

AC

+

GND

3. Block DC signal while passing AC signal.

4. Bypass AC signal around a circuit or to ground.

5. Filter out unwanted portions of a fluctuating signal.

6. Use with resistor to integrate a fluctuating signal:

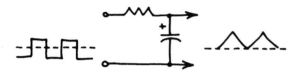

7. Or to differentiate a fluctuating signal:

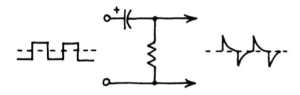

8. Perform a timing function:

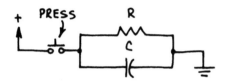

C will quickly charge...then slowly discharge through R.

9. Store a charge to keep a transistor turned off or on.

10. Store a charge to be dumped through a flashtube or LED in a fast and powerful pulse.

Can you substitute capacitors? In most cases changing the value of a capacitor 10% or even 100% will not cause a malfunction, but circuit operation may be affected. In a timing circuit, for example, increasing the value of the timing capacitor will increase the timing period. Changing the capacitors in a filter will change the filter's frequency response. Be sure to use the proper voltage rating. And don't worry about the difference between 0.47 and 0.5 μF.

SEMICONDUCTORS

Usually made from silicon. Be sure to observe all operating restrictions. Brief descriptions of important semiconductor devices:

DIODES

Permit current to flow in but one direction (forward bias). Used to rectify AC, allow current to flow into a circuit but block its return, etc.

ZENER DIODES

The zener diode is a voltage regulator. In this typical circuit, voltage exceeding the diode's breakdown voltage is shunted to ground:

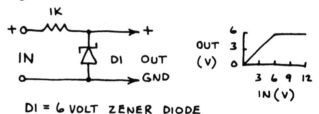

DI = 6 VOLT ZENER DIODE

Zeners can also protect voltage sensitive components and provide a convenient reference voltage.

LIGHT EMITTING DIODES

LEDs emit green, yellow, red or infrared when forward biased. A series resistor should be used to limit current to less than the maximum allowed:

$$R_S = \frac{V_{CC} - V_{LED}}{LED_I}$$

Example: V_{LED} of red LED is 1.7 volts. For a forward current (LED_I) of 20 mA at V_{CC} = 5 volts, R = 165 ohms. Don't exceed LED_I!!

3

Infrared LEDs are much more powerful than visible LEDs, but their radiation is totally invisible. Use them for object detectors and communicators.

TRANSISTORS

In this notebook, transistors are used as simple amplifiers and switches that turn on LEDs. Any general purpose switching transistors will work.

INTEGRATED CIRCUITS

Since an IC is a complete circuit on a silicon chip, you must observe all operating restrictions. Reversed polarity, excessive supply voltage and sourcing or sinking too much current can destroy an IC. Be sure to pay close attention to the location of the power supply pins! Most ICs are packaged in 8, 14 or 16 pin plastic DIPs (Dual In-line Packages). A notch or circle is near pin 1:

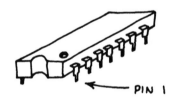

PIN 1

When the IC is right side up, pin 1 is at lower left:

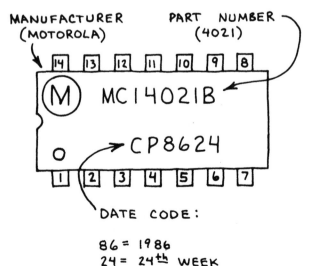

MANUFACTURER (MOTOROLA)   PART NUMBER (4021)

DATE CODE:

86 = 1986
24 = 24th WEEK

Incidentally, a date code may not be present, but other numbers may be...and the date code is not always below the device number:

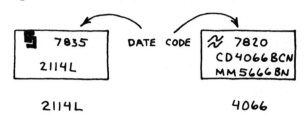

DATE CODE

2114L   4066

Store ICs in a plastic cabinet if you can afford one. Or insert them in rows in a styrofoam tray (the kind used for meat in a grocery store). CAUTION: Never store MOS/CMOS ICs in ordinary non-conductive plastic.

DIGITAL INTEGRATED CIRCUITS

INTRODUCTION

DIGITAL ICs ARE 2-STATE DEVICES. ONE STATE IS NEAR 0 VOLTS OR GROUND (LOW OR L) AND THE OTHER IS NEAR THE IC'S SUPPLY VOLTAGE (HIGH OR H). SUBSTITUTE 0 FOR L AND 1 FOR H AND DIGITAL ICs CAN PROCESS INDIVIDUAL BINARY DIGITS (BITS) OR MULTIPLE BIT WORDS. A 4-BIT WORD IS A <u>NIBBLE</u> AND AN 8-BIT WORD IS A <u>BYTE</u>.

THE BINARY SYSTEM

IT'S VERY HELPFUL TO KNOW THE FIRST 16 BINARY NUMBERS. IF 0=L AND 1=H, THEY ARE:

0 - L L L L		8 - H L L L
1 - L L L H		9 - H L L H
2 - L L H L		10 - H L H L
3 - L L H H		11 - H L H H
4 - L H L L		12 - H H L L
5 - L H L H		13 - H H L H
6 - L H H L		14 - H H H L
7 - L H H H		15 - H H H H

NOTE THAT LLLL (0) IS AS MUCH A NUMBER AS ANY OTHER NUMBER.

LOGIC GATES

LOGIC CIRCUITS ARE MADE BY INTER-CONNECTING TWO OR MORE OF THESE BASIC LOGIC GATES:

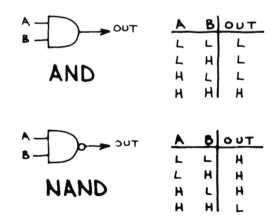

AND

A	B	OUT
L	L	L
L	H	L
H	L	L
H	H	H

NAND

A	B	OUT
L	L	H
L	H	H
H	L	H
H	H	L

OR

A	B	OUT
L	L	L
L	H	H
H	L	H
H	H	H

NOR

A	B	OUT
L	L	H
L	H	L
H	L	L
H	H	L

EXCLUSIVE-OR

A	B	OUT
L	L	L
L	H	H
H	L	H
H	H	L

EXCLUSIVE-NOR

A	B	OUT
L	L	H
L	H	L
H	L	L
H	H	H

YES (BUFFER)

A	OUT
L	L
H	H

NOT (INVERTER)

A	OUT
L	H
H	L

3-STATE LOGIC

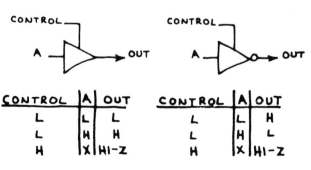

CONTROL	A	OUT
L	L	L
L	H	H
H	X	HI-Z

CONTROL	A	OUT
L	L	H
L	H	L
H	X	HI-Z

HI-Z: OUTPUT IN HIGH IMPEDANCE STATE.

MOS/CMOS INTEGRATED CIRCUITS

INTRODUCTION

MOS ICs CAN CONTAIN MORE FUNC-
TIONS PER CHIP THAN TTL/LS AND
ARE VERY EASY TO USE. MOST CHIPS
IN THIS SECTION ARE CMOS (COM-
PLEMENTARY MOS). THEY CONSUME VERY
LITTLE POWER AND OPERATE OVER A
+3-15 VOLT RANGE. CMOS CAN BE POW-
ERED BY THIS:

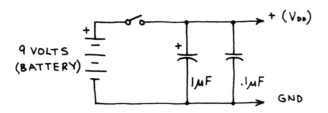

OR YOU CAN USE A LINE POWERED
SUPPLY MADE FROM A 7805/7812/7815.
SEE THE LINEAR SECTION.

INCIDENTALLY, YOU CAN POWER A
CMOS CIRCUIT FROM TWO SERIES
CONNECTED PENLIGHT CELLS, BUT
A 9-12 VOLT SUPPLY WILL GIVE
BETTER PERFORMANCE.

OPERATING REQUIREMENTS

1. THE INPUT VOLTAGE SHOULD <u>NOT</u>
EXCEED V_{DD}! (TWO EXCEPTIONS:
THE 4049 AND 4050.)

2. AVOID, IF POSSIBLE, SLOWLY RISING
AND FALLING INPUT SIGNALS SINCE
THEY CAN CAUSE EXCESSIVE POWER
CONSUMPTION. RISETIMES FASTER
THAN 15 MICROSECONDS ARE BEST.

3. <u>ALL</u> UNUSED INPUTS <u>MUST</u> BE
CONNECTED TO V_{DD} (+) OR V_{SS} (GND).
OTHERWISE ERRATIC CHIP BEHAVIOR
AND EXCESSIVE CURRENT CONSUMPTION
WILL OCCUR.

4. <u>NEVER</u> CONNECT AN INPUT
SIGNAL TO A CMOS CIRCUIT WHEN
THE POWER IS OFF.

5. OBSERVE HANDLING PRECAUTIONS.

6

HANDLING PRECAUTIONS

A CMOS CHIP IS MADE FROM PMOS
AND NMOS TRANSISTORS. MOS MEANS
<u>M</u>ETAL - <u>O</u>XIDE - <u>S</u>ILICON (OR <u>S</u>EMICONDUCTOR).
P AND N REFER TO POSITIVE AND
NEGATIVE CHANNEL MOS TRANSISTORS.
AN NMOS TRANSISTOR LOOKS LIKE THIS:

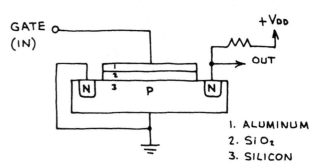

1. ALUMINUM
2. SiO_2
3. SILICON

A PMOS TRANSISTOR IS IDENTICAL
EXCEPT THE P AND N REGIONS ARE
EXCHANGED. THE SiO_2 (SILICON DIOXIDE)
LAYER IS A GLASSY FILM THAT
SEPARATES AND INSULATES THE METAL
GATE FROM THE SILICON SUBSTRATE.
THIS FILM IS WHY A MOS TRANSISTOR
OR IC PLACES PRACTICALLY NO LOAD
ON THE SOURCE OF AN INPUT SIGNAL.
THE FILM IS VERY THIN AND IS THERE-
FORE EASILY PUNCTURED BY STATIC
ELECTRICITY:

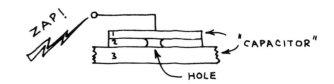

PREVENT STATIC DISCHARGE!

1. <u>NEVER</u> STORE MOS IC's IN NONCON-
DUCTIVE PLASTIC "SNOW," TRAYS, BAGS
OR FOAM.

2. PLACE MOS IC's <u>PINS DOWN</u> ON AN
ALUMINUM FOIL SHEET OR TRAY WHEN
THEY ARE <u>NOT</u> IN A CIRCUIT OR
STORED IN CONDUCTIVE FOAM.

3. USE A BATTERY POWERED IRON TO
SOLDER MOS CHIPS. DO NOT USE AN
AC POWERED IRON.

INTERFACING CMOS

1. IF SUPPLY VOLTAGES ARE EQUAL:

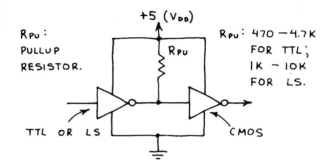

R_{PU}: PULLUP RESISTOR.

R_{PU}: 470 — 4.7K FOR TTL; 1K — 10K FOR LS.

TTL OR LS → CMOS

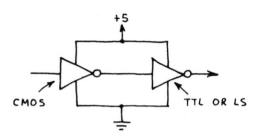

CMOS → TTL OR LS

2. DIFFERENT SUPPLY VOLTAGES:

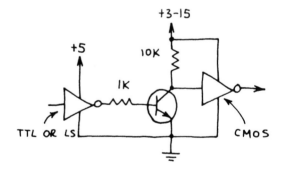

TTL OR LS → CMOS

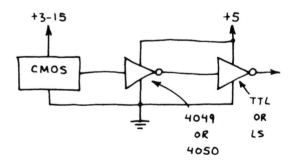

CMOS → 4049 OR 4050 → TTL OR LS

NOTE THAT CMOS MUST BE POWERED BY AT LEAST 5 VOLTS WHEN CMOS IS INTERFACED WITH TTL. OTHERWISE THE CMOS INPUT WILL EXCEED V_{DD}.

3. CMOS LED DRIVERS:

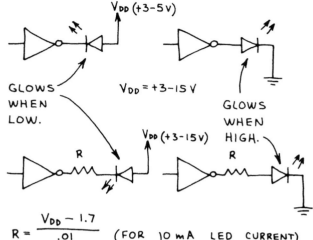

GLOWS WHEN LOW.

$V_{DD} = +3-15 V$

GLOWS WHEN HIGH.

$$R = \frac{V_{DD} - 1.7}{.01} \quad \text{(FOR 10 mA LED CURRENT)}$$

USE 1000 OHMS FOR MOST APPLICATIONS.

CMOS LOGIC CLOCK

MANY CIRCUITS IN THIS SECTION REQUIRE A SOURCE OF PULSES. HERE'S A SIMPLE CMOS CLOCK:

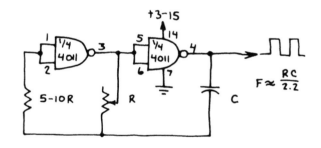

$$F \approx \frac{RC}{2.2}$$

TYPICAL VALUES: R = 100K, C = 0.01 — 0.1 μF

OK TO USE 4049 ... BUT MUCH MORE CURRENT WILL BE REQUIRED.

CMOS TROUBLESHOOTING

1. DO **ALL** INPUTS GO SOMEWHERE?

2. ARE **ALL** IC PINS INSERTED INTO THE BOARD OR SOCKET?

3. IS THE IC HOT? IF SO, SEE 1-2 ABOVE AND MAKE SURE THE OUTPUT IS NOT OVERLOADED.

4. DOES THE CIRCUIT OBEY **ALL** CMOS OPERATING REQUIREMENTS?

5. HAVE YOU FORGOTTEN A CONNECTION?

7

QUAD NAND GATE
4011

THE BASIC CMOS BUILDING BLOCK CHIP. MORE APPLICATIONS THAN TTL 7400/74LS00 QUAD NAND GATE.

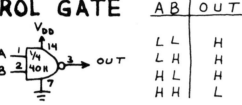

V_{DD} (+3-15V)

IMPORTANT: CONNECT ALL UNUSED INPUTS TO PIN 7 OR 14!

CONTROL GATE

A	B	OUT
L	L	H
L	H	H
H	L	H
H	H	L

INVERTER

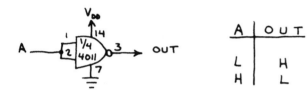

A	OUT
L	H
H	L

AND GATE

A	B	OUT
L	L	L
L	H	L
H	L	L
H	H	H

OR GATE

A	B	OUT
L	L	L
L	H	H
H	L	H
H	H	H

AND-OR GATE

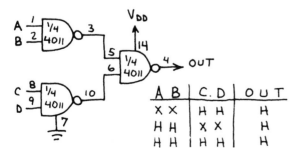

A	B	C	D	OUT
X	X	H	H	H
H	H	X	X	H
H	H	H	H	H

NOR GATE

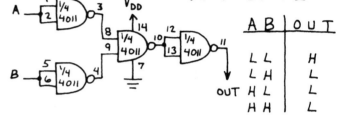

A	B	OUT
L	L	H
L	H	L
H	L	L
H	H	L

4-INPUT NAND GATE

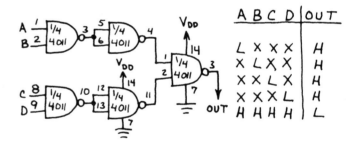

A	B	C	D	OUT
L	X	X	X	H
X	L	X	X	H
X	X	L	X	H
X	X	X	L	H
H	H	H	H	L

EXCLUSIVE-OR GATE

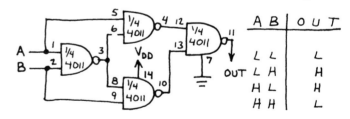

A	B	OUT
L	L	L
L	H	H
H	L	H
H	H	L

EXCLUSIVE-NOR GATE

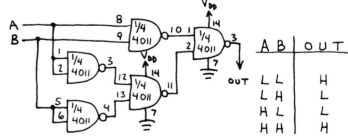

A	B	OUT
L	L	H
L	H	L
H	L	L
H	H	H

QUAD NAND GATE (CONTINUED)
4011

GATED OSCILLATOR

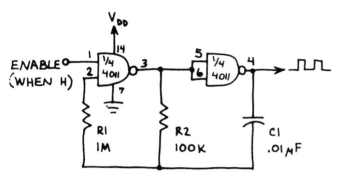

OUTPUT FREQUENCY IS
1 KHz SQUARE WAVE.

TOUCH SWITCH

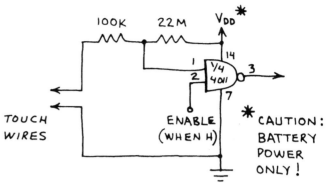

OUTPUT GOES HIGH WHEN
TOUCH WIRES ARE BRIDGED
BY A FINGER.

SIMPLE OSCILLATOR

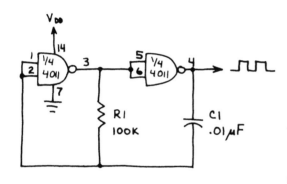

OUTPUT NOT AS SYMMETRICAL
AS ABOVE CIRCUIT.

ONE-SHOT TOUCH SWITCH

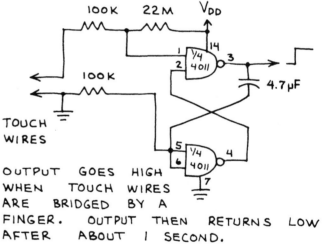

OUTPUT GOES HIGH
WHEN TOUCH WIRES
ARE BRIDGED BY A
FINGER. OUTPUT THEN RETURNS LOW
AFTER ABOUT 1 SECOND.

GATED FLASHER

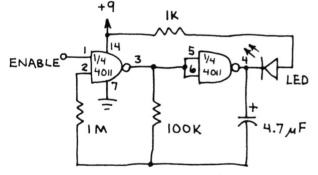

LED FLASHES 1-2 Hz
WHEN ENABLE IS HIGH.
LED STAYS ON WHEN
ENABLE IS LOW.

INCREASED OUTPUT DRIVE

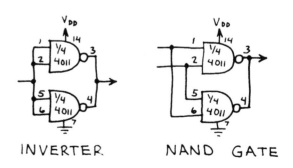

INVERTER   NAND GATE

USE THIS METHOD TO INCREASE
CURRENT THE 4011 CAN SOURCE
OR SINK. OK TO ADD MORE GATES.

9

QUAD NOR GATE
4001

AN IMPORTANT CMOS BUILDING
BLOCK CHIP. ITS HIGH IMPEDANCE
INPUT MAKES POSSIBLE MORE
APPLICATIONS THAN THE TTL 7402/
74LS02 QUAD NOR GATE.

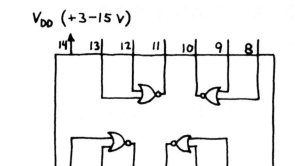

V_{DD} (+3-15 V)

IMPORTANT: CONNECT ALL UNUSED
INPUTS TO PIN 7 OR 14.

BOUNCELESS SWITCH

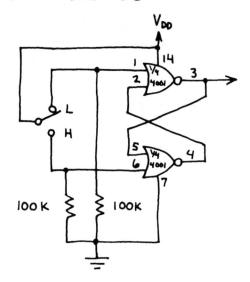

100K   100K

INCREASED OUTPUT DRIVE

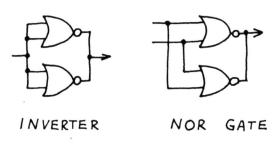

INVERTER   NOR GATE

USE THIS METHOD TO INCREASE
CURRENT THE 4001 CAN SOURCE
OR SINK. OK TO ADD MORE GATES.

GATED TONE SOURCE

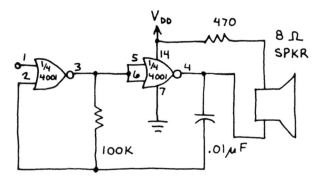

TONE FREQUENCY IS ABOUT 1KHz.

LED FLASHER

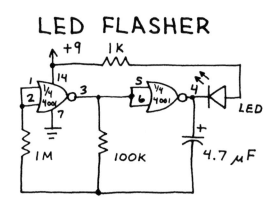

LED FLASHES 1-2 TIMES/SECOND.

RS LATCH

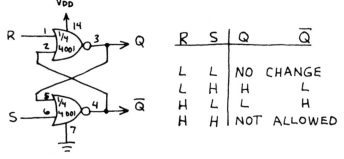

R	S	Q	Q̄
L	L	NO CHANGE	
L	H	H	L
H	L	L	H
H	H	NOT ALLOWED	

OR GATE

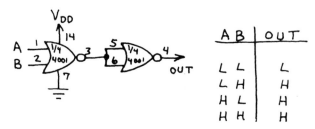

A	B	OUT
L	L	L
L	H	H
H	L	H
H	H	H

10

QUAD AND GATE
4081

BUILDING BLOCK CHIP. USE
FOR BUFFERING AND LOGIC.
NOT AS VERSATILE AS 4011.

$V_{DD} (+3-15V)$

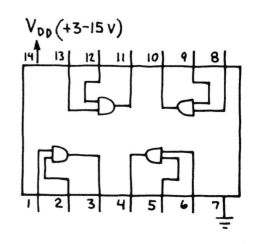

AND GATE BUFFER

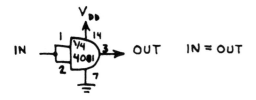

IN = OUT

DIGITAL TRANSMISSION GATE

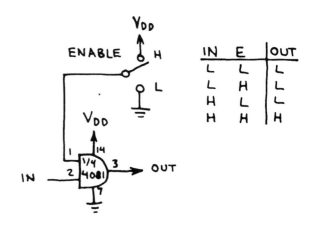

IN	E	OUT
L	L	L
L	H	L
H	L	L
H	H	H

NAND GATE

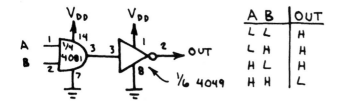

A B	OUT
L L	H
L H	H
H L	H
H H	L

1/6 4049

NOR GATE

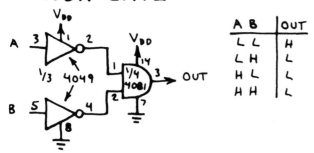

A B	OUT
L L	H
L H	L
H L	L
H H	L

AND-OR-INVERT GATE

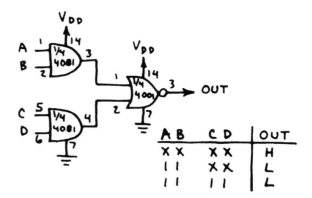

A B	C D	OUT
X X	X X	H
1 1	X X	L
1 1	1 1	L

4-INPUT NAND GATE

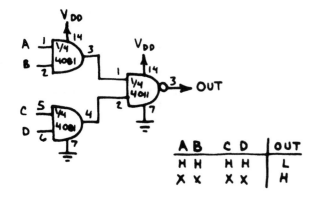

A B	C D	OUT
H H	H H	L
X X	X X	H

4-INPUT AND GATE

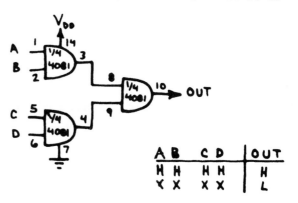

A B	C D	OUT
H H	H H	H
X X	X X	L

11

QUAD EXCLUSIVE-OR GATE 4070

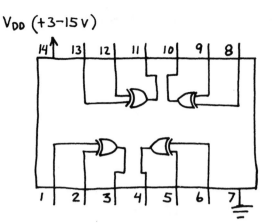

V_{DD} (+3-15V)

THE OUTPUT OF EACH GATE GOES LOW WHEN BOTH INPUTS ARE EQUAL. THE OUTPUT GOES HIGH IF THE INPUTS ARE UNEQUAL. MANY APPLICATIONS INCLUDING BINARY ADDITION, COMPARING BINARY WORDS AND PHASE DETECTION.

IMPORTANT: CONNECT UNUSED INPUTS TO PIN 7 OR 14.

CONTROLLED INVERTER

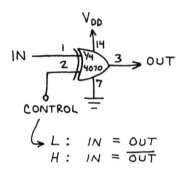

L: IN = OUT
H: IN = $\overline{OUT}$

1-BIT COMPARATOR

THIS CIRCUIT IS ALSO A HALF-ADDER WITHOUT A CARRY OUTPUT.

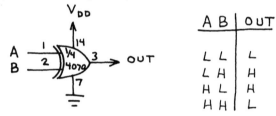

A B	OUT
L L	L
L H	H
H L	H
H H	L

BINARY FULL ADDER

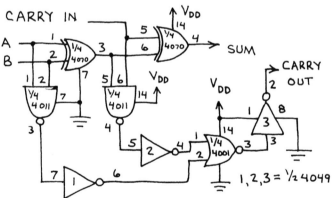

1,2,3 = ½ 4049

4-BIT COMPARATOR

DETERMINES IF TWO 4-BIT WORDS ARE EQUAL.

HINT: USE 4011 (P.8-9) IF 4012 IS UNAVAILABLE.

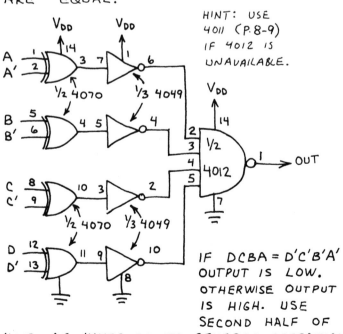

IF DCBA = D'C'B'A' OUTPUT IS LOW. OTHERWISE OUTPUT IS HIGH. USE SECOND HALF OF 4012 AS INVERTER TO REVERSE OPERATION.

PHASE DETECTOR

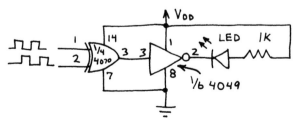

LED STOPS GLOWING WHEN THE INPUT FREQUENCIES ARE EQUAL.

12

QUAD EXCLUSIVE-OR GATE (CONTINUED)
4070

EXCLUSIVE-NOR

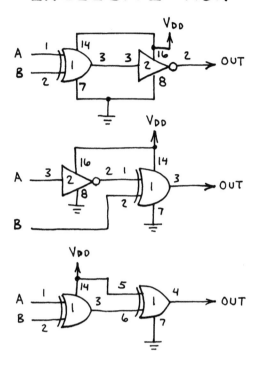

IC1 = ¼ 4070
IC2 = ⅙ 4049

A B	OUT
L L	H
L H	L
H L	L
H H	H

8-INPUT EX-OR

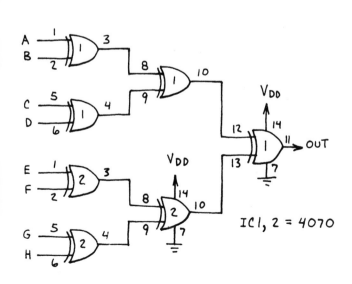

IC1, 2 = 4070

3-INPUT EX-OR

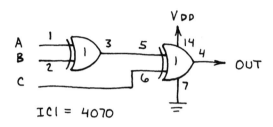

IC1 = 4070

10 MHz OSCILLATOR

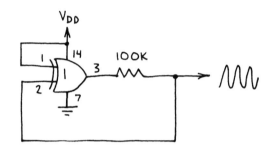

V_{DD} = 3 TO 15 VOLTS

FREQUENCY VARIES WITH V_{DD}:

V_{DD}	FREQUENCY	AMPLITUDE
5	2.4 MHz	3.5 V
10	9.4 MHz	8.0 V
15	11.0 MHz	12.0 V

SQUARE WAVE GENERATOR

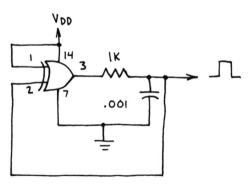

V_{DD} = 3 TO 15 VOLTS

RISETIME = 50 NANOSECONDS
FREQUENCY = 2 MHz WHEN
V_{DD} = 10 VOLTS

13

HEX INVERTING BUFFER
4049

IN ADDITION TO STANDARD
LOGIC AND CMOS TO TTL
INTERFACING, OFTEN USED
IN OSCILLATORS AND PULSE
GENERATORS. FOR LOW CURRENT
APPLICATIONS, USE 4011 CONNECTED
AS INVERTER. (OK TO USE 4011 FOR
CIRCUITS ON THIS PAGE.)

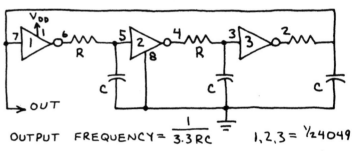

VDD (+3-15V)

NOTE UNUSUAL LOCATION
OF POWER SUPPLY PINS.

CLOCK PULSE GENERATOR

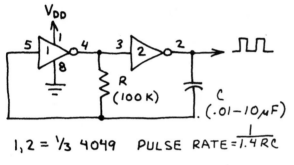

1,2 = 1/3 4049   PULSE RATE = $\frac{1}{1.4RC}$

PHASE SHIFT OSCILLATOR

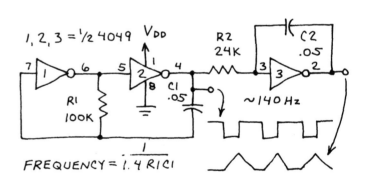

OUTPUT FREQUENCY = $\frac{1}{3.3RC}$ 1,2,3 = 1/2 4049

BOUNCELESS SWITCH

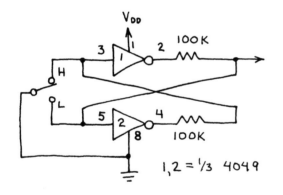

1,2 = 1/3 4049

TRIANGLE WAVE SOURCE

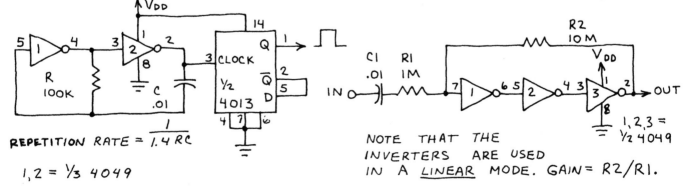

1,2,3 = 1/2 4049

R2 24K   C2 .05

~140Hz

FREQUENCY = $\frac{1}{1.4R1C1}$

SQUARE WAVE GENERATOR

R 100K   C .01

3 CLOCK   1 Q
1/2 4013   Q̄ 2
D 5

REPETITION RATE = $\frac{1}{1.4RC}$

1,2 = 1/3 4049

LINEAR 10X AMPLIFIER

R2 10M

C1 .01   R1 1M

IN   OUT

NOTE THAT THE
INVERTERS ARE USED
IN A <u>LINEAR</u> MODE. GAIN = R2/R1.

1,2,3 = 1/2 4049

14

HEX NON-INVERTING BUFFER
4050

PRIMARILY INTENDED FOR
INTERFACING CMOS TO TTL.
SUPPLIES MORE CURRENT
THAN STANDARD CMOS.

IMPORTANT: ALL UNUSED INPUTS
MUST GO TO PIN 1 OR 8.

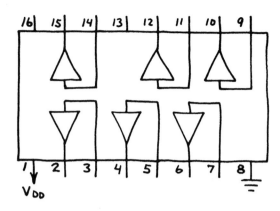

NOTE UNUSUAL LOCATION
OF POWER SUPPLY PINS.

OUTPUT EXPANDER

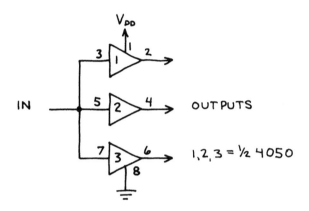

IN

OUTPUTS

1,2,3 = ½ 4050

LOGIC PROBE

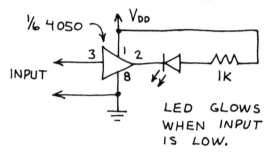

LED GLOWS
WHEN INPUT
IS LOW.

OUTPUT BUFFER

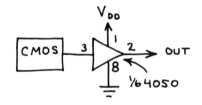

INCREASED OUTPUT DRIVE

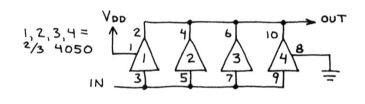

1,2,3,4 =
⅔ 4050

CMOS TO CMOS
AT LOWER V_DD

CMOS TO TTL/LS
AT LOWER V_cc

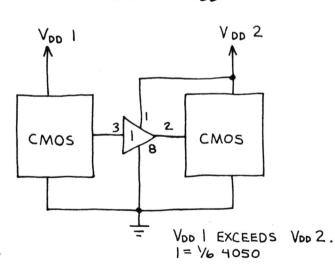

V_DD 1 EXCEEDS V_DD 2.
1 = ½ 4050

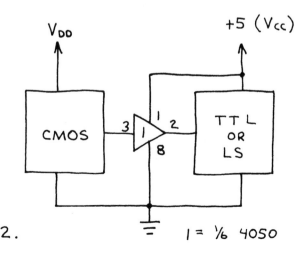

1 = ⅙ 4050

15

DUAL 4-INPUT NAND GATE 4012

VERY USEFUL IN MAKING DECODERS. ALSO CAN BE USED TO ADD ONE OR MORE ENABLE INPUTS TO VARIOUS CIRCUITS.

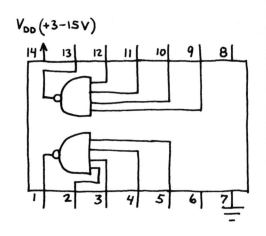

V_{DD} (+3-15V)

ENABLE INPUT

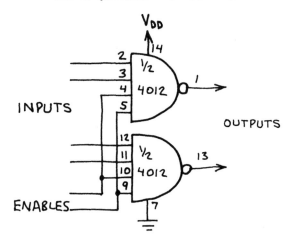

INPUTS

ENABLES

OUTPUTS

BCD DECODERS

DECIMAL 0

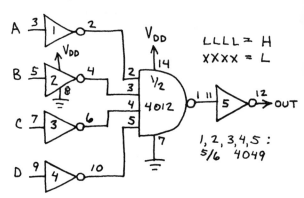

LLLL = H
XXXX = L

1, 2, 3, 4, 5:
5/6 4049

OUT

1-OF-4 DECODER

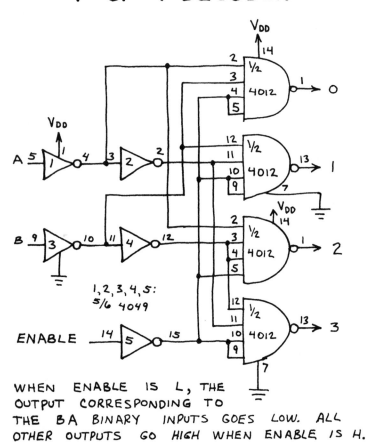

1, 2, 3, 4, 5:
5/6 4049

ENABLE

WHEN ENABLE IS L, THE OUTPUT CORRESPONDING TO THE BA BINARY INPUTS GOES LOW. ALL OTHER OUTPUTS GO HIGH WHEN ENABLE IS H.

DECIMAL 1

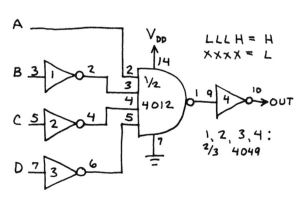

LLLH = H
XXXX = L

OUT

1, 2, 3, 4:
2/3 4049

DECIMAL 9

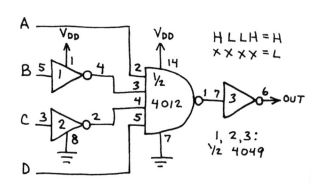

HLLH = H
XXXX = L

OUT

1, 2, 3:
1/2 4049

16

TRIPLE 3-INPUT NAND GATE
4023

HANDY FOR MAKING CUSTOM DECODERS,
CONVERTERS AND MULTIPLE INPUT GATES.

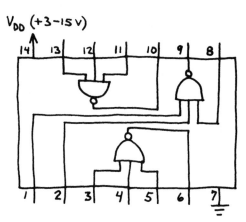

V_{DD} (+3-15V)

6-INPUT OR GATE

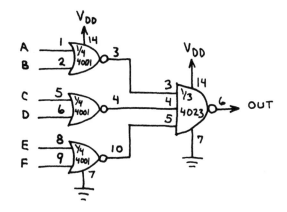

IMPORTANT: CONNECT ALL UNUSED
INPUTS TO PIN 7 OR 14.

9-INPUT NAND GATE

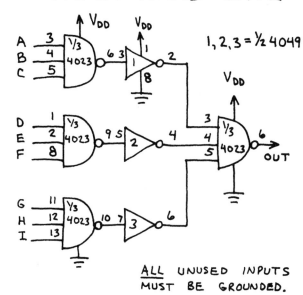

1,2,3 = ½ 4049

ALL UNUSED INPUTS
MUST BE GROUNDED.

DECIMAL-TO-BCD CONVERTER

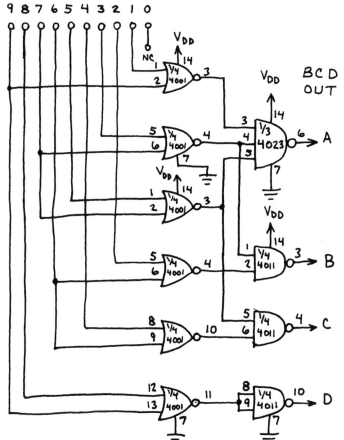

DECIMAL IN (SELECTED DIGIT H,
ALL OTHERS L.)

1-OF-4 DECODER

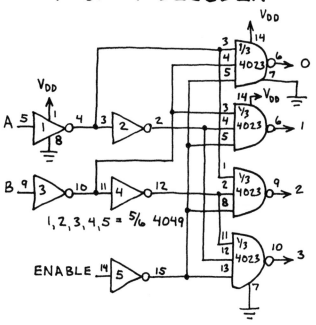

1,2,3,4,5 = 5/6 4049

ENABLE

17

QUAD BILATERAL SWITCH
4066

ONE OF THE MOST VERSATILE
CMOS CHIPS. PINS A, B, C AND D
CONTROL FOUR ANALOG SWITCHES.
CLOSE A SWITCH BY CONNECTING
ITS CONTROL PIN TO V_{DD}. ON
RESISTANCE = 80−250 OHMS.
OPEN A SWITCH BY CONNECTING ITS
CONTROL PIN TO GROUND (PIN 7).
OFF RESISTANCE = 10^9 OHMS. I/O (INPUT/
OUTPUT) AND O/I PINS ARE REVERSIBLE.

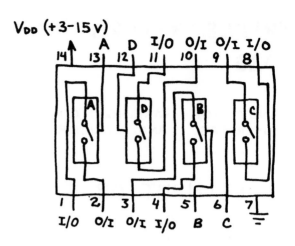

DATA BUS CONTROL

DATA SELECTOR

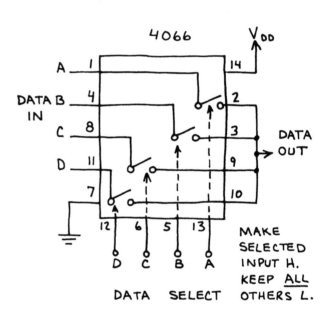

CONTROL:
L = OFF
H = LOAD

MAKE
SELECTED
INPUT H.
KEEP ALL
OTHERS L.

DATA SELECT

DIGITAL TO ANALOG (D/A) CONVERTER

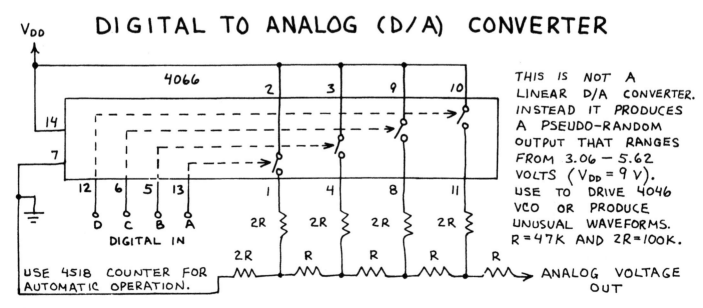

THIS IS NOT A
LINEAR D/A CONVERTER.
INSTEAD IT PRODUCES
A PSEUDO-RANDOM
OUTPUT THAT RANGES
FROM 3.06 − 5.62
VOLTS (V_{DD} = 9 V).
USE TO DRIVE 4046
VCO OR PRODUCE
UNUSUAL WAVEFORMS.
R = 47K AND 2R = 100K.

USE 4518 COUNTER FOR
AUTOMATIC OPERATION.

18

PROGRAMMABLE GAIN AMPLIFIER

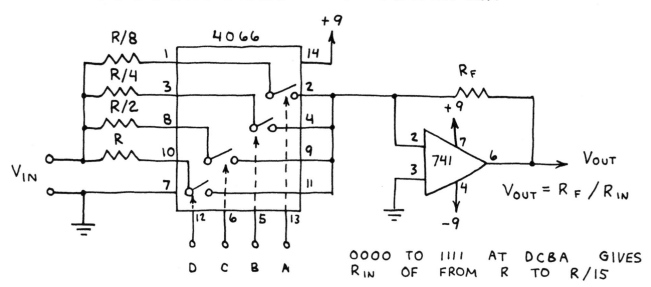

$$V_{OUT} = R_F / R_{IN}$$

0000 TO 1111 AT DCBA GIVES R_{IN} OF FROM R TO R/15

PROGRAMMABLE FUNCTION GENERATOR

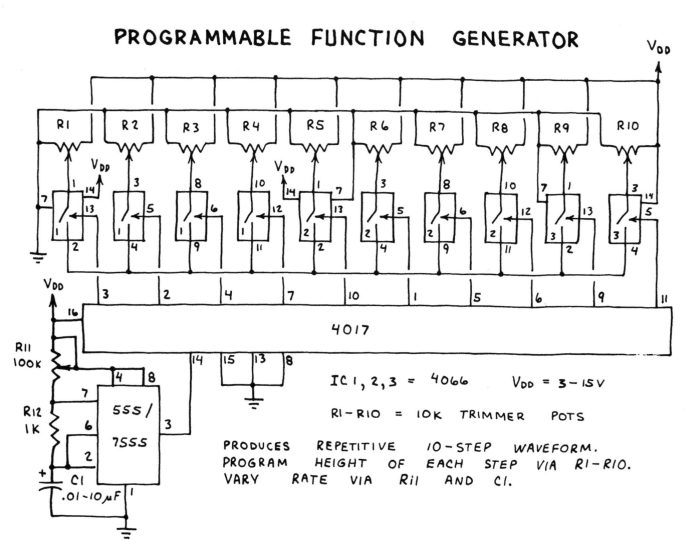

IC 1, 2, 3 = 4066 V_{DD} = 3 - 15V

R1 - R10 = 10K TRIMMER POTS

PRODUCES REPETITIVE 10-STEP WAVEFORM.
PROGRAM HEIGHT OF EACH STEP VIA R1-R10.
VARY RATE VIA R11 AND C1.

1024-BIT STATIC RAM
2102L

1024 1-BIT STORAGE LOCATIONS ADDRESSED BY PINS A0-A9. TTL/LS COMPATIBLE. CE (CHIP ENABLE) INPUT CONTROLS R/W (READ/WRITE) OPERATIONS). 3-STATE OUTPUTS.

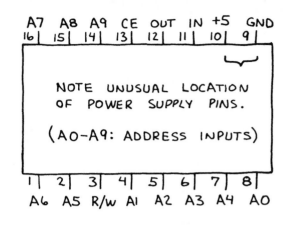

CE	R/W	OPERATION
L	L	WRITE (LOADS BIT AT PIN 11)
L	H	READ (OUTPUTS BIT AT PIN 12)
H	X	HI Z (OUTPUT ENTERS THIRD STATE)

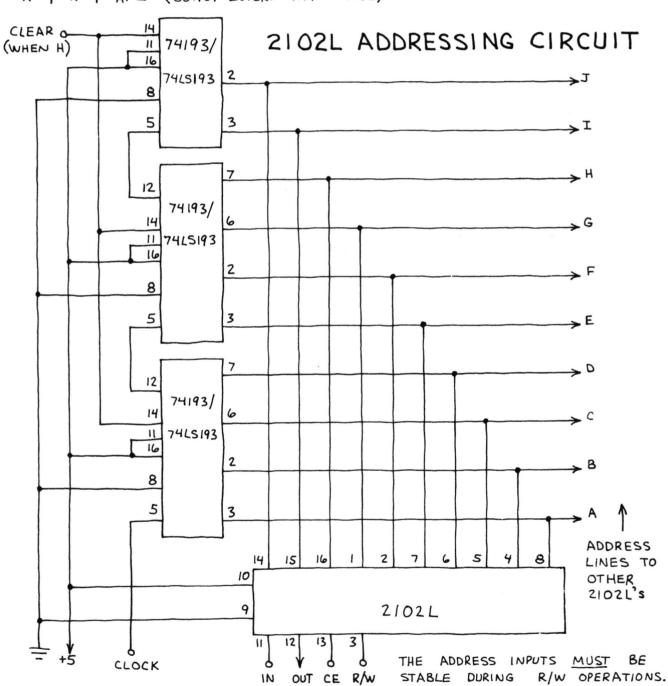

2102L ADDRESSING CIRCUIT

THE ADDRESS INPUTS MUST BE STABLE DURING R/W OPERATIONS.

ADDING PROGRAMMED OR MANUAL JUMP

ADD THESE CONNECTIONS TO THE ADDRESSING CIRCUIT ON FACING PAGE.

SA—SJ: USE 8-POSITION DIP SWITCHES OR MINIATURE TOGGLES. OPEN=H ; CLOSED=L

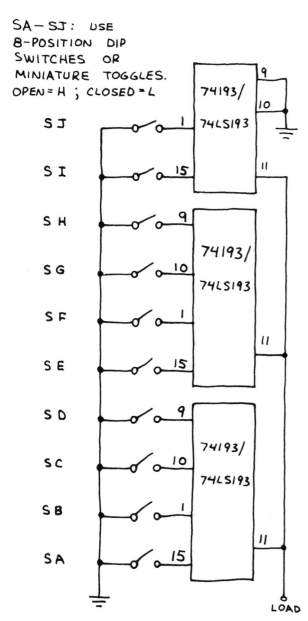

NORMALLY THE LOAD INPUT IS HIGH. MAKING LOAD LOW LOADS THE ADDRESS PROGRAMMED IN SWITCHES SA-SJ INTO THE 74193's. THIS PERMITS A PROGRAMMED JUMP OR A MANUAL JUMP TO ANY ADDRESS.

SINGLE I/O PORT

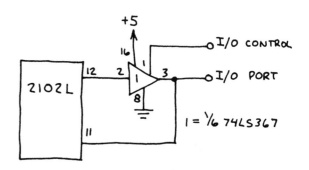

$1 = \frac{1}{6}$ 74LS367

ADD THIS CIRCUIT TO THE ADDRESSING CIRCUIT ON FACING PAGE. WHEN I/O (INPUT/OUTPUT) CONTROL IS H, PIN 3 OF THE 74LS367 ENTERS THIRD STATE (HI-Z) AND I/O PORT ACCEPTS INPUT DATA. WHEN PIN 3 OF THE 74LS367 IS L, I/O PORT OUTPUTS DATA. BOTH THESE OPERATIONS ARE DEPENDENT UPON THE STATUS OF THE 2102L CONTROL INPUTS.

CASCADING 2102L'S

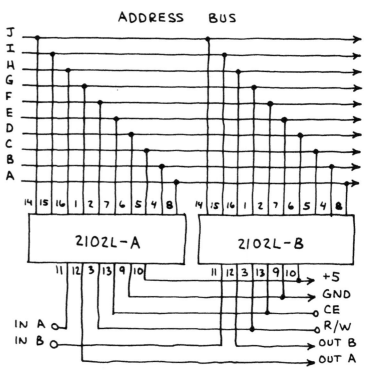

1024 × 4-BIT RAM
2114L /4045

1024-4-BIT STORAGE LOCATIONS ADDRESSED
BY PINS A0-A9. TTL/LS COMPATIBLE.
FOR READ/WRITE OPERATIONS, CE (CHIP ENABLE,
ALSO CALLED CHIP SELECT) MUST BE LOW.
WE INPUT MUST BE LOW TO WRITE
(LOAD) DATA INTO CHIP.   WHEN WE
IS HIGH, DATA IN ADDRESSED
LOCATION APPEARS AT INPUT/OUTPUT
PINS. IDEAL CHIP FOR DO-IT-YOURSELF
MICROCOMPUTERS AND CONTROLLERS.

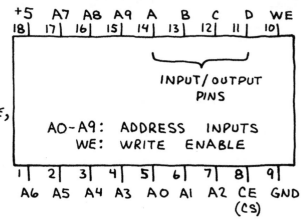

INPUT/OUTPUT PINS

AO-A9: ADDRESS INPUTS
WE: WRITE ENABLE

2114L ADDRESSING CIRCUIT

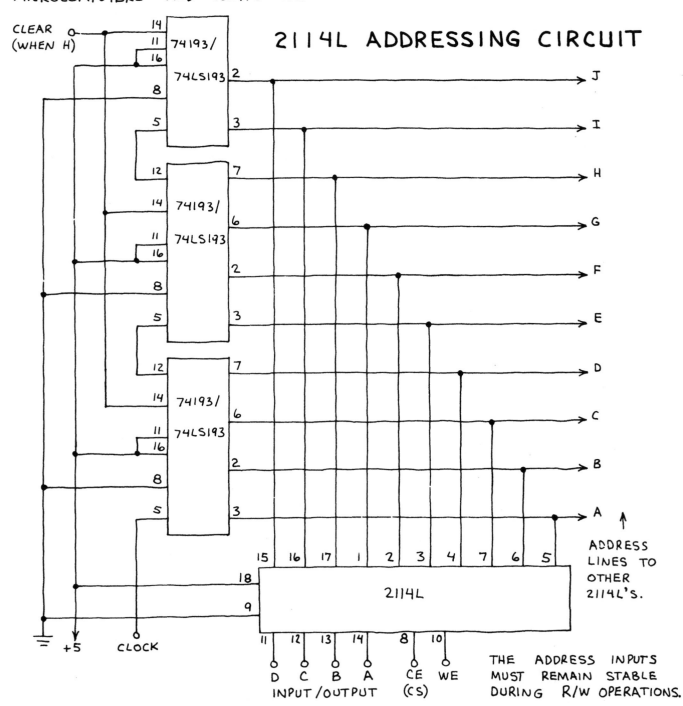

THE ADDRESS INPUTS
MUST REMAIN STABLE
DURING R/W OPERATIONS.

1024 × 4-BIT RAM (CONTINUED)
2114L/4045

1024-NIBBLE
DATA LOADING CIRCUIT

(NIBBLE = 4-BIT WORD OR ½ 8-BIT WORD)

MANUAL JUMP: 1. SET SWITCHES A-J
TO DESIRED ADDRESS; 2. PRESS S6.

USE THIS CIRCUIT TO MANUALLY
STORE UP TO 1024 4-BIT WORDS
IN A 2114L. AFTER THE DATA
IS LOADED, IT CAN THEN BE READ
BACK AT THE CLOCK SPEED. THE
DATA OUTPUTS ARE PINS 11-14 WHEN
DATA INPUT SWITCHES ARE AT NEUTRAL.

WRITE: 1. SWITCH S2 TO THE
BOUNCELESS PUSHBUTTON.
2. SWITCH S4 AND S5 TO L.
3. CLOSE S3.
4. INPUT DATA.
5. PRESS BOUNCELESS PUSHBUTTON.
6. REPEAT STEPS 1-5.

READ: 1. OPEN S3.
2. SWITCH S5 TO H.
3. CLOSE, THEN OPEN, S1.
4. SELECT CLOCKED OR
MANUAL OUTPUT (S2).

NOTE:
BEST TO OUTPUT DATA
THROUGH 74LS367 HEX
BUFFER.

S1 — CLEAR

J I

SPST TOGGLES

H G F E

.1 µF

USE AT ALL IC PWR SPY PINS.

D C B A

S6 — LOAD

(NORMALLY CLOSED)

74193/74LS193
74193/74LS193
74193/74LS193

S2
SPDT TOGGLES
DPST TOGGLE

CLOCK
BOUNCELESS PUSHBUTTON

S3 — WRITE

+5

2114L

CE (CS)

D   C   B   A   WE

S4 — CHIP ENABLE
L H

DATA INPUT SWITCHES
(SPDT WITH NEUTRAL CENTER)

S5 — WRITE ENABLE
L H
+5

23

DUAL D FLIP-FLOP 4013

VERY VERSATILE PAIR OF D-TYPE
FLIP-FLOPS. GROUND UNUSED INPUTS.

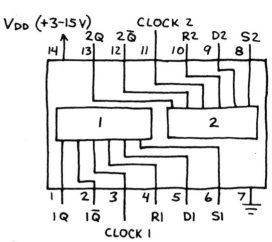

1-OF-4 SEQUENCER

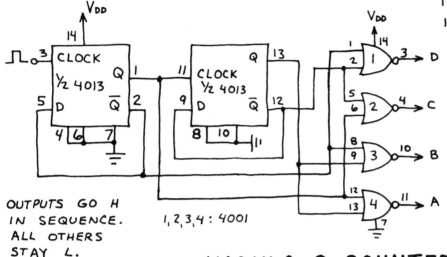

OUTPUTS GO H
IN SEQUENCE.
ALL OTHERS
STAY L.

1, 2, 3, 4 : 4001

DIVIDE-BY-2

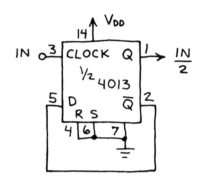

MODULO-8 COUNTER

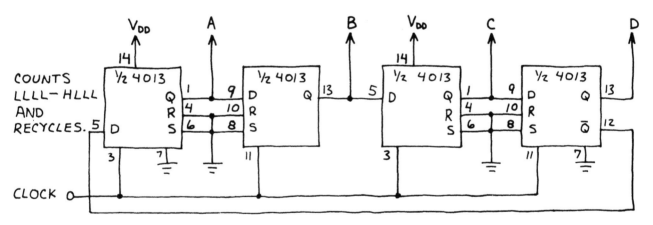

COUNTS
LLLL-HLLL
AND
RECYCLES.

SERIAL IN/OUT, PARALLEL OUT SHIFT REGISTER

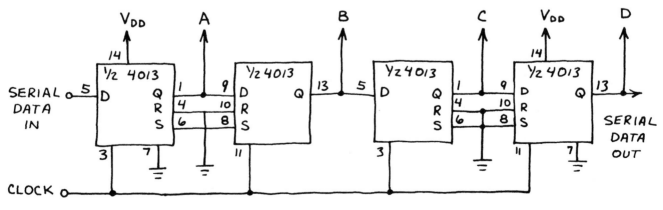

SERIAL
DATA
IN

SERIAL
DATA
OUT

24

DUAL JK FLIP FLOP
4027

USE FOR DIVIDERS, COUNTERS AND REGISTERS. S (SET) AND R (RESET) INPUTS MUST BE LOW FOR CLOCKING TO OCCUR. MAKING S OR R HIGH SETS OR RESETS FLIP-FLOP INDEPENDENT OF CLOCK. IMPORTANT: ALL INPUTS MUST GO SOMEWHERE!

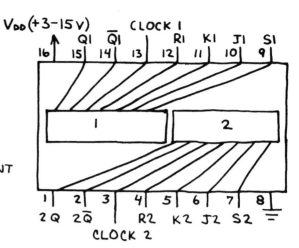

DIVIDE-BY-2 COUNTER

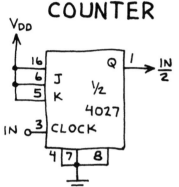

DIVIDE-BY-5 COUNTER

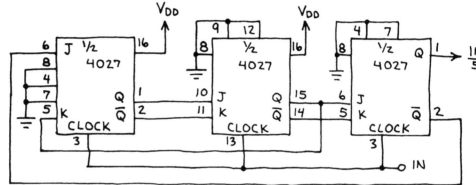

DIVIDE-BY-3 COUNTER

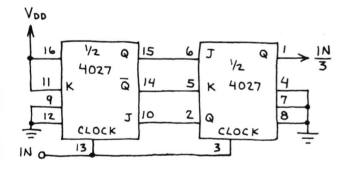

DIVIDE-BY-4 COUNTER

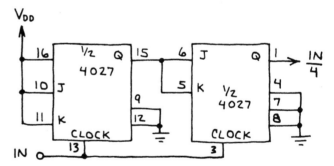

4-BIT SERIAL SHIFT REGISTER

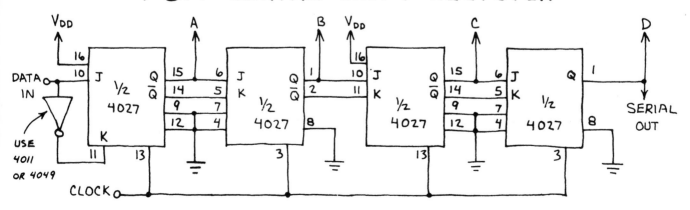

25

QUAD LATCH
4042

FOUR BISTABLE LATCHES.
CAN BE USED AS A
4-BIT DATA REGISTER.
ALL FOUR LATCHES ARE
CLOCKED SIMULTANEOUSLY.
POLARITY PIN PROVIDES
CLOCKING FLEXIBILITY.

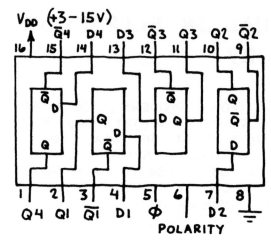

4-BIT DATA LATCH

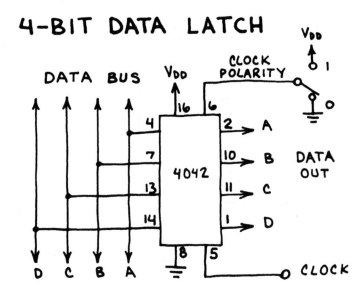

CLOCK	POLARITY	Q
0	0	D
⌐_	0	LATCH
_⌐	1	D
	1	LATCH

DATA ON BUS APPEARS
AT OUTPUTS. DATA
IS LATCHED (SAVED)
WHEN CLOCK SWITCHES.

STEPPED WAVE GENERATOR

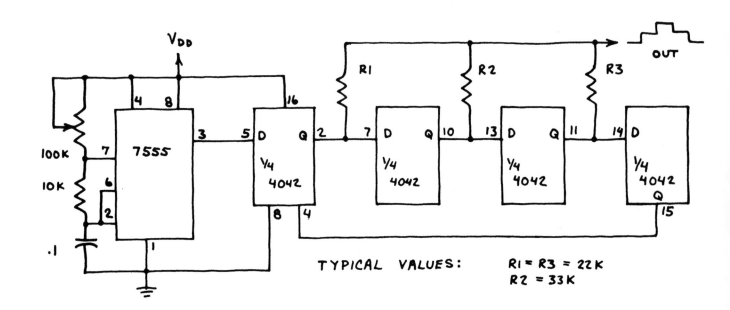

TYPICAL VALUES: R1 = R3 = 22K
                   R2 = 33K

26

DUAL ONE-SHOT
4528

TWO FULLY INDEPENDENT
MONOSTABLE MULTIVIBRATORS.
BOTH CAN BE RETRIGGERED.
TRIGGER CAN BE RISING
OR FALLING EDGE OF PULSE.
T1 AND T2 ARE TIMING INPUTS.
RST IS RESET AND $\pm$IN ARE
TRIGGER INPUTS.

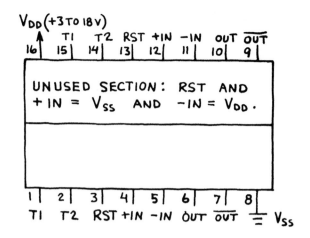

POSITIVE ONE-SHOT

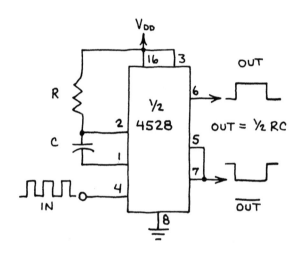

OUT = ½ RC

PULSE DELAYER

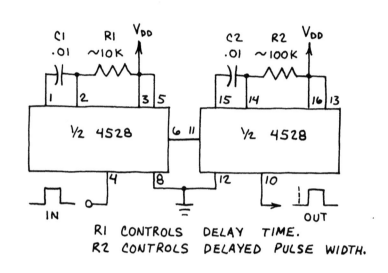

R1 CONTROLS DELAY TIME.
R2 CONTROLS DELAYED PULSE WIDTH.

STEPPED TONE GENERATOR

TO CONTROL
WITH LIGHT,
USE CdS
PHOTOCELL
FOR R1.

ADJUST R1 TO CREATE
UNIQUE STEPPED TONE.
R2 CONTROLS FREQUENCY.
OK TO EXPERIMENT
WITH C1 AND C2.
R3 CONTROLS GAIN.

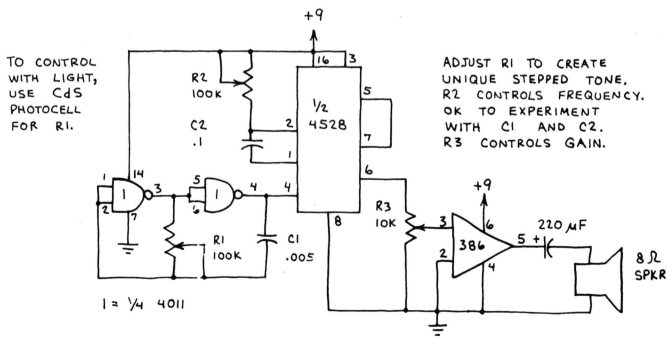

27

14-STAGE BINARY COUNTER 4020

A RIPPLE COUNTER WITH CARRY OUTPUT. THE 14-STAGE BINARY COUNT IS COMPLETED IN 16,384 CLOCK PULSES. THIS MAKES POSSIBLE VERY LONG DURATION TIMERS, ASSUMING THE OUTPUTS ARE DECODED. THE OUTPUTS REQUIRE A BRIEF SETTLING TIME AFTER EACH CLOCK PULSE.

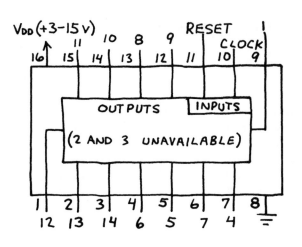

14-BIT BINARY COUNTER

THE SECOND AND THIRD OUTPUTS (÷4 AND ÷8) OF THE 4020 ARE <u>NOT</u> AVAILABLE. THIS CIRCUIT INCLUDES A 3-BIT COUNTER TO SUPPLY THE MISSING OUTPUTS. <u>A</u> IS THE LOWEST ORDER OUTPUT.

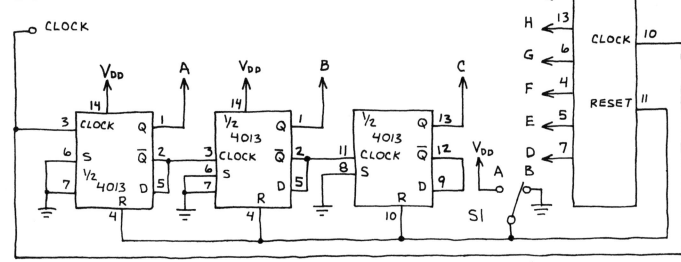

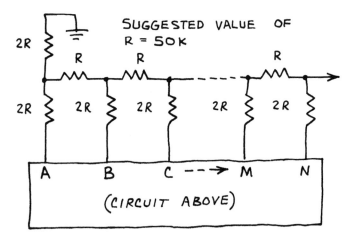

SUGGESTED VALUE OF R = 50K

(CIRCUIT ABOVE)

STAIRCASE GENERATOR

ANALOG OUTPUT

OUTPUT IS A STEPPED VOLTAGE. APPLICATIONS INCLUDE ANALOG-TO-DIGITAL CONVERSION AND WAVEFORM SYNTHESIS.

DUAL BCD COUNTER
4518

TWO SYNCHRONOUS DECADE
COUNTERS IN ONE PACKAGE.
WHEN ENABLE IS HIGH AND
RESET IS LOW, EACH COUNTER
ADVANCES ONE COUNT PER
CLOCK PULSE.

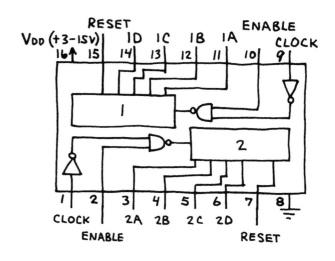

CASCADED BCD COUNTERS

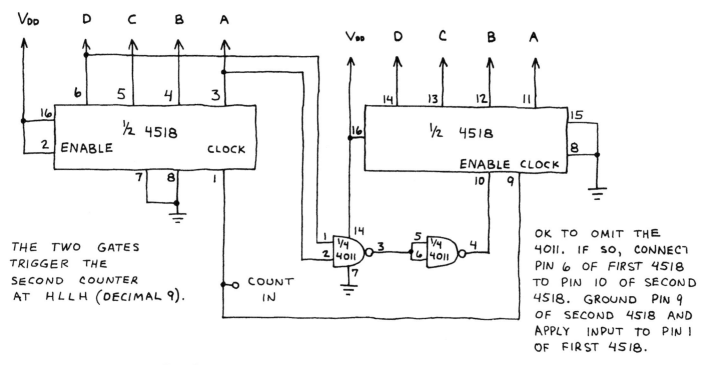

THE TWO GATES
TRIGGER THE
SECOND COUNTER
AT HLLH (DECIMAL 9).

OK TO OMIT THE
4011. IF SO, CONNECT
PIN 6 OF FIRST 4518
TO PIN 10 OF SECOND
4518. GROUND PIN 9
OF SECOND 4518 AND
APPLY INPUT TO PIN 1
OF FIRST 4518.

BCD KEYBOARD ENCODER

PRESS S0-S9, THEN TOGGLE RESET SWITCH
S10 TO V_DD AND BACK TO GROUND.
BCD EQUIVALENT OF SELECTED KEY (S0-S9) APPEARS →

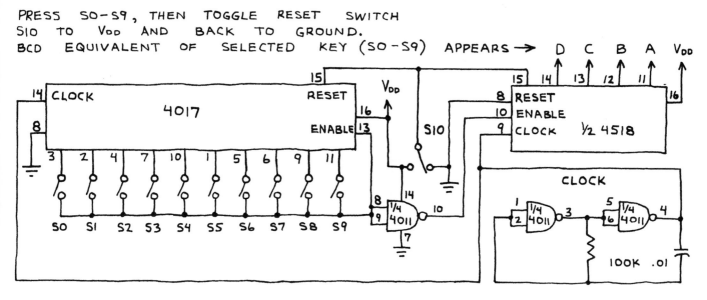

29

DECADE COUNTER/DIVIDER 4017

SEQUENTIALLY MAKES 1-OF-10 OUTPUTS HIGH (OTHERS STAY LOW) IN RESPONSE TO CLOCK PULSES. MANY APPLICATIONS. COUNT TAKES PLACE WHEN PINS 13 AND 15 ARE LOW.

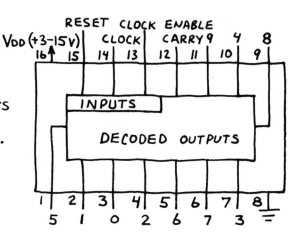

RANDOM NUMBER GENERATOR

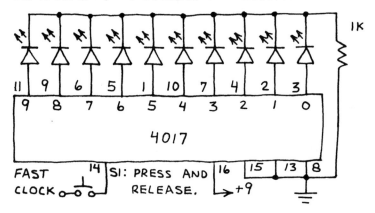

COUNT TO N AND HALT

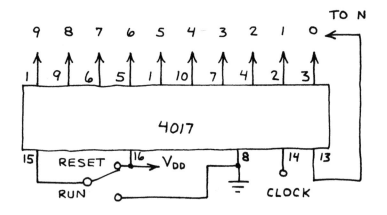

COUNT TO N AND RECYCLE

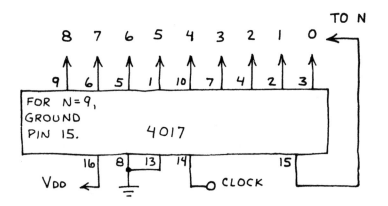

0-99 COUNTER

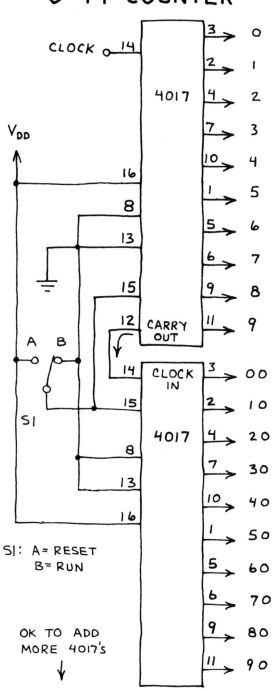

DECADE COUNTER/DIVIDER (CONTINUED)
4017

BCD KEYBOARD ENCODER

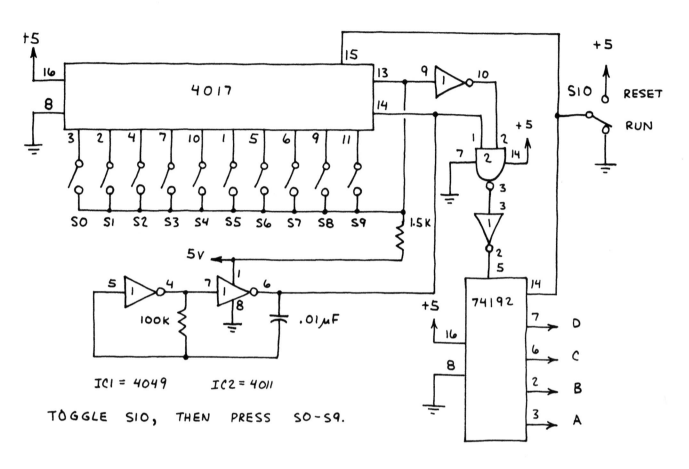

ICI = 4049   IC2 = 4011

TOGGLE S10, THEN PRESS S0-S9.

FREQUENCY DIVIDER

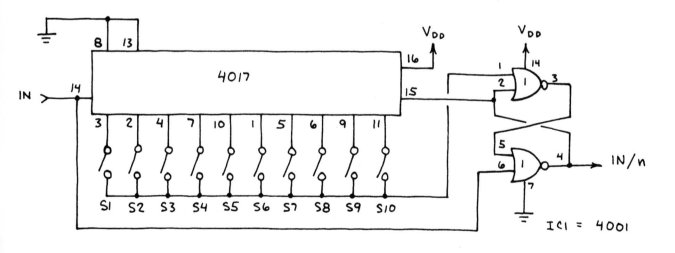

ICI = 4001

CLOSE S1-S10 TO DIVIDE
FREQUENCY BY FROM 1 TO 10.

3-DIGIT BCD COUNTER MC14553

COMPLETE 3-DIGIT COUNTER. USE FOR
DO-IT-YOURSELF EVENT AND FREQUENCY
COUNTERS. BEGINNERS: GET SOME
PRACTICAL CIRCUIT EXPERIENCE BEFORE
USING THIS CHIP. PIN EXPLANATIONS:
DS (DIGIT SELECT) 1,2,3— SEQUENTIALLY
STROBES READOUTS. LE—LATCH ENABLE
(WHEN H). DIS— INHIBITS INPUT WHEN H.
CLOCK—INPUT. MR— MASTER RESET (WHEN H).
OF —OVERFLOW. A,B,C,D — BCD OUTPUTS.

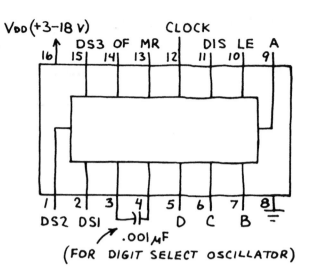

(FOR DIGIT SELECT OSCILLATOR)

3-DIGIT EVENT COUNTER

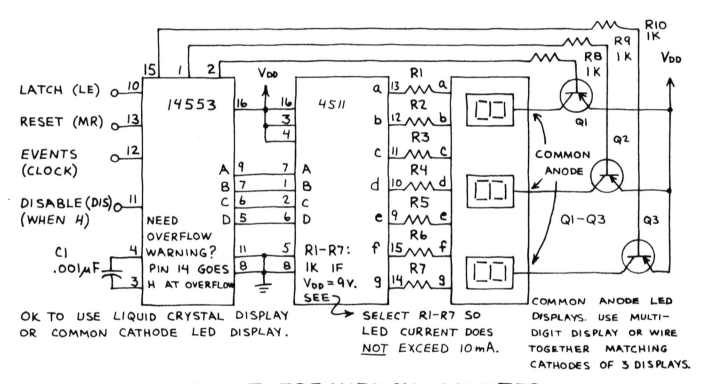

OK TO USE LIQUID CRYSTAL DISPLAY
OR COMMON CATHODE LED DISPLAY.

SELECT R1-R7 SO
LED CURRENT DOES
NOT EXCEED 10mA.

COMMON ANODE LED
DISPLAYS. USE MULTI-
DIGIT DISPLAY OR WIRE
TOGETHER MATCHING
CATHODES OF 3 DISPLAYS.

6-DIGIT FREQUENCY COUNTER

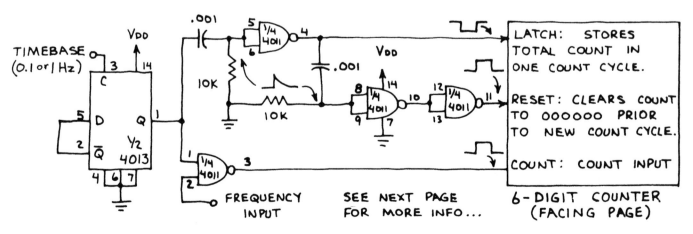

LATCH: STORES
TOTAL COUNT IN
ONE COUNT CYCLE.

RESET: CLEARS COUNT
TO 000000 PRIOR
TO NEW COUNT CYCLE.

COUNT: COUNT INPUT

SEE NEXT PAGE
FOR MORE INFO...

6-DIGIT COUNTER
(FACING PAGE)

3-DIGIT BCD COUNTER (CONTINUED)
MC14553

6-DIGIT COUNTER

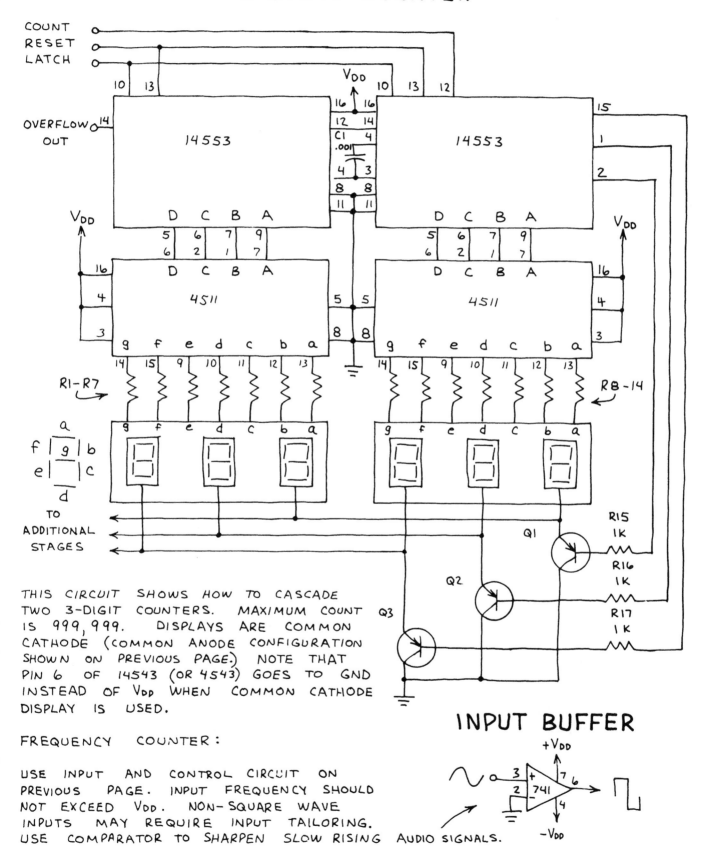

THIS CIRCUIT SHOWS HOW TO CASCADE TWO 3-DIGIT COUNTERS. MAXIMUM COUNT IS 999,999. DISPLAYS ARE COMMON CATHODE (COMMON ANODE CONFIGURATION SHOWN ON PREVIOUS PAGE.) NOTE THAT PIN 6 OF 14543 (OR 4543) GOES TO GND INSTEAD OF V_{DD} WHEN COMMON CATHODE DISPLAY IS USED.

FREQUENCY COUNTER:

USE INPUT AND CONTROL CIRCUIT ON PREVIOUS PAGE. INPUT FREQUENCY SHOULD NOT EXCEED V_{DD}. NON-SQUARE WAVE INPUTS MAY REQUIRE INPUT TAILORING. USE COMPARATOR TO SHARPEN SLOW RISING AUDIO SIGNALS.

INPUT BUFFER

BCD-TO-DECIMAL DECODER 4028

DECODES 4-BIT BCD INPUT INTO 1-OF-10 OUTPUTS. SELECTED OUTPUT GOES HIGH; ALL OTHERS STAY LOW. USE FOR DECIMAL READOUTS, SEQUENCERS, PRO-GRAMMABLE COUNTERS, ETC.

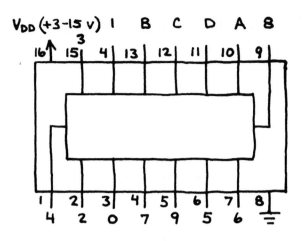

0-9 SECOND TIMER

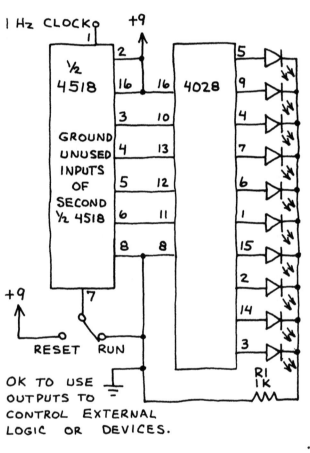

OK TO USE OUTPUTS TO CONTROL EXTERNAL LOGIC OR DEVICES.

1-OF-8 DECODER

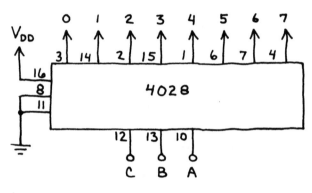

ADDRESS INPUTS

COUNT TO N AND HALT

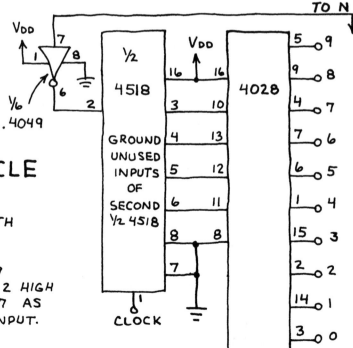

COUNT TO N AND RECYCLE

USE THE ADJACENT CIRCUIT WITH THESE CHANGES:

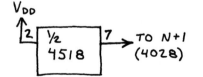

1. OMIT 4049
2. MAKE PIN 2 HIGH
3. USE PIN 7 AS CONTROL INPUT.

BCD-TO-7-SEGMENT LATCH/DECODER/DRIVER 4511

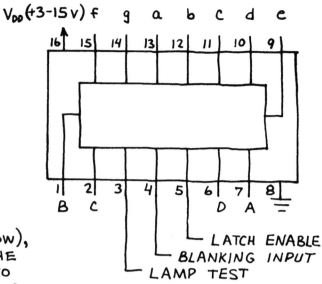

CONVERTS BCD DATA INTO FORMAT SUITABLE FOR PRODUCING DECIMAL DIGITS ON 7-SEGMENT LED DISPLAY. INCLUDES BUILT-IN 4-BIT LATCH TO STORE DATA TO BE DISPLAYED (WHEN PIN 5 IS HIGH). WHEN LATCH IS NOT USED (PIN 5 LOW), THE 7-SEGMENT OUTPUTS FOLLOW THE BCD INPUTS. MAKE PIN 4 LOW TO EXTINGUISH THE DISPLAY AND HIGH FOR NORMAL OPERATION. MAKE PIN 3 LOW TO TEST THE DISPLAY AND HIGH FOR NORMAL OPERATION.

DISPLAY FLASHER

DISPLAY FLASHES ONCE PER SECOND WHEN E IS HIGH.

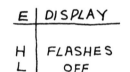

E	DISPLAY
H	FLASHES
L	OFF

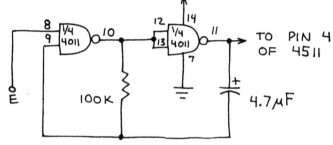

DECIMAL COUNTING UNIT (DCU)

IMPORTANT: ALL INPUTS MUST GO SOMEWHERE!

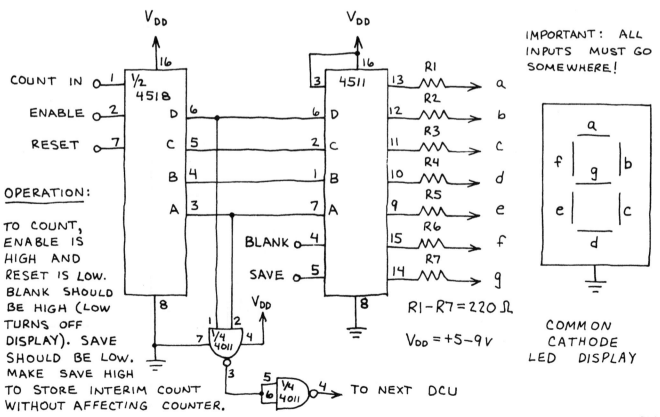

OPERATION:

TO COUNT, ENABLE IS HIGH AND RESET IS LOW. BLANK SHOULD BE HIGH (LOW TURNS OFF DISPLAY). SAVE SHOULD BE LOW. MAKE SAVE HIGH TO STORE INTERIM COUNT WITHOUT AFFECTING COUNTER.

R1-R7 = 220 Ω

V_{DD} = +5-9V

COMMON CATHODE LED DISPLAY

8-STAGE SHIFT REGISTER
4021

PARALLEL INPUT / SERIAL OUTPUT SHIFT REGISTER. ALSO SERIAL INPUT. DATA AT PARALLEL INPUTS IS FORCED INTO THE REGISTER IRRESPECTIVE OF THE CLOCK STATUS WHEN PIN 9 IS MADE HIGH. KEEP PIN 9 LOW FOR NORMAL OPERATION.

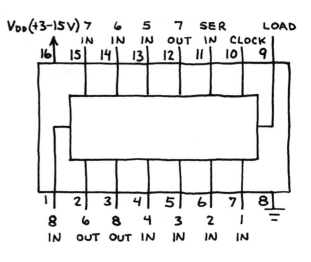

PARALLEL-TO-SERIAL DATA CONVERTER

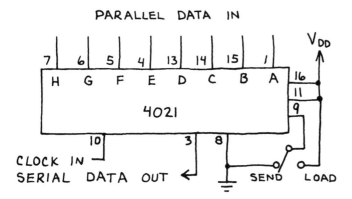

ALL 1's (H's) ARE SENT AFTER THE 8-BIT WORD IS TRANSMITTED.

8-STAGE DELAY LINE

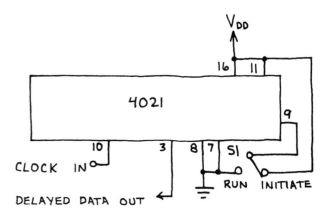

THE FIRST PARALLEL INPUT (PIN 7) IS GROUNDED. THIS LOADS A SINGLE L WHEN SI IS SWITCHED TO INITIATE. THE SINGLE L BIT REACHES THE OUTPUT AFTER 8 CLOCK PULSES.

PSEUDO-RANDOM SEQUENCER

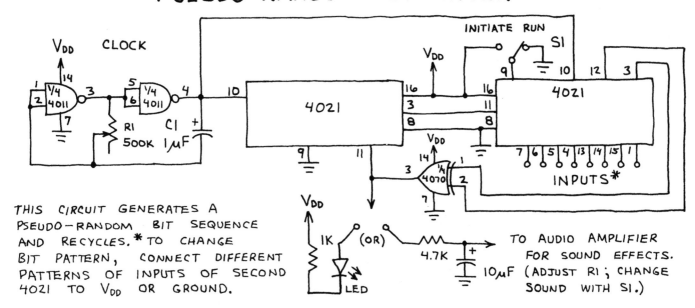

THIS CIRCUIT GENERATES A PSEUDO-RANDOM BIT SEQUENCE AND RECYCLES. *TO CHANGE BIT PATTERN, CONNECT DIFFERENT PATTERNS OF INPUTS OF SECOND 4021 TO VDD OR GROUND.

TO AUDIO AMPLIFIER FOR SOUND EFFECTS. 10μF (ADJUST R1; CHANGE SOUND WITH SI.)

ANALOG MULTIPLEXER 4051

INPUT ADDRESS AT CBA SELECTS 1-OF-8 ANALOG SWITCHES. SIGNAL AT SELECTED SWITCH I/O (INPUT/ OUTPUT) IS THEN APPLIED TO COMMON O/I (OUTPUT/INPUT). THE INPUT SIGNAL MUST NOT EXCEED V_{DD}. THE INHIBIT (INH) INPUT SHOULD BE GROUNDED FOR NORMAL OPERATION. ALL SWITCHES ARE OPEN WHEN INH IS HIGH.

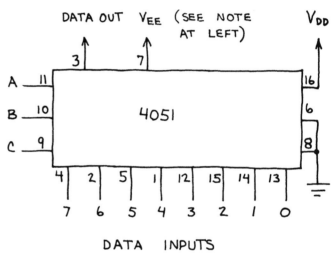

1-OF-8 MULTIPLEXER

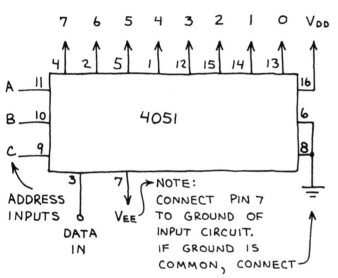

ADDRESS INPUTS

DATA IN

V_{EE}

NOTE: CONNECT PIN 7 TO GROUND OF INPUT CIRCUIT. IF GROUND IS COMMON, CONNECT

1-OF-8 DATA SELECTOR (DEMULTIPLEXER)

DATA INPUTS

TONE SEQUENCER

CYCLES THROUGH 8 TONES AND REPEATS. R1 CONTROLS TEMPO. R2-R9 ARE INDIVIDUAL TONE RESISTORS. USE 1K-100K EACH.

DO NOT REDUCE R10. USE AMPLIFIER FOR MORE VOLUME.

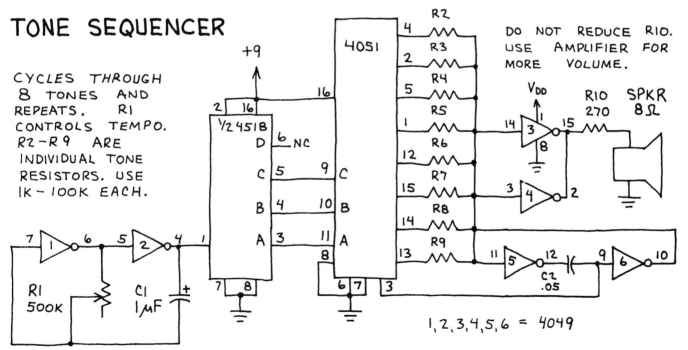

1,2,3,4,5,6 = 4049

37

60-Hz TIMEBASE
MM5369

PROVIDES PRECISE 60 Hz SQUARE WAVE
WHEN USED WITH 3.579545 MHz
COLOR TV CRYSTAL. USE FOR MOST
DO-IT-YOURSELF TIMERS, CLOCKS, CONTROLLERS,
FUNCTION GENERATORS. INSTALL IN SMALL
CABINET FOR WORKBENCH PRECISION CLOCK.

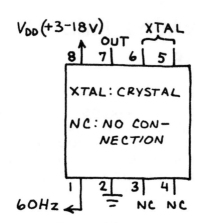

60-Hz TIMEBASE

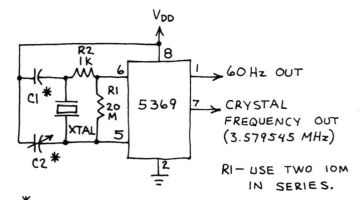

60 Hz OUT

CRYSTAL
FREQUENCY OUT
(3.579545 MHz)

R1— USE TWO 10M
IN SERIES.

* MOTOROLA SPECIFIES THAT C1 = 30pF
AND C2 = 6.36 pF. OK TO USE SIX
4.7 pF CAPACITORS IN PARALLEL OR
47pF CAPACITOR FOR C1. TRY TUNABLE
CAPACITOR (e.g. 5-50pF) FOR C2. TO
TUNE, CONNECT FREQUENCY METER
TO PIN 7. TUNE C2 UNTIL FREQUENCY
IS 3,579,545 Hz. ACCURACY FAIRLY
GOOD EVEN IF YOU DON'T TUNE C2.

10-Hz TIMEBASE

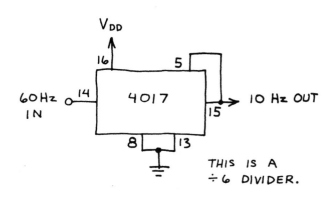

60Hz IN

10 Hz OUT

THIS IS A
÷6 DIVIDER.

1-Hz TIMEBASE

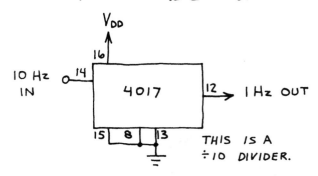

10 Hz IN

1 Hz OUT

THIS IS A
÷10 DIVIDER.

DIGITAL STOPWATCH

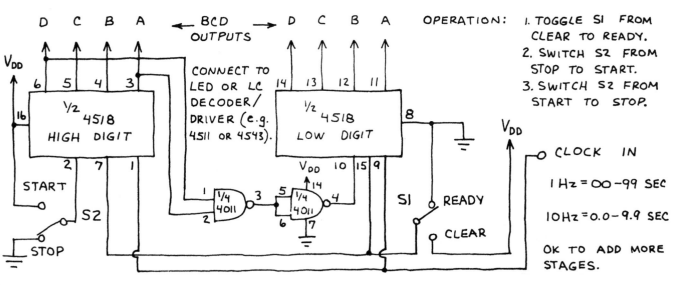

OPERATION: 1. TOGGLE S1 FROM
CLEAR TO READY.
2. SWITCH S2 FROM
STOP TO START.
3. SWITCH S2 FROM
START TO STOP.

CLOCK IN

1 Hz = 00-99 SEC

10Hz = 0.0-9.9 SEC

OK TO ADD MORE
STAGES.

NOISE GENERATOR
S2688 / MM5837N

PRODUCES BROADBAND WHITE NOISE FOR AUDIO AND OTHER APPLICATIONS. THE NOISE QUALITY IS VERY UNIFORM. IT IS PRODUCED BY A 17-BIT SHIFT REGISTER WHICH IS CLOCKED BY AN INTERNAL OSCILLATOR.

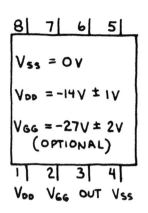

$V_{SS} = 0V$

$V_{DD} = -14V \pm 1V$

$V_{GG} = -27V \pm 2V$ (OPTIONAL)

V_DD V_GG OUT V_SS

WHITE NOISE SOURCE

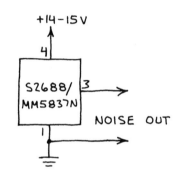

NOISE OUT

CONNECT OUTPUT TO AUDIO AMPLIFIER TO HEAR NOISE. USE 7815 VOLTAGE REGULATOR TO OBTAIN +15 VOLTS.

PINK NOISE SOURCE

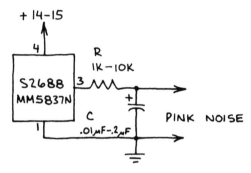

PINK NOISE

CHANGE R AND C TO ALTER NOISE SPECTRUM. ALSO, TRY LOWER SUPPLY VOLTAGES TO CHANGE SPECTRUM.

COIN TOSSER

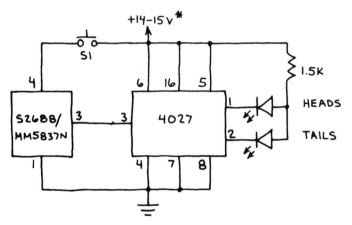

HEADS

TAILS

PRESS S1; BOTH LEDS GLOW. RELEASE S1 AND ONLY ONE GLOWS. GROUND INPUTS OF UNUSED HALF OF 4027 (PINS 9,10,11,12 AND 13).*(OK TO USE 9-VOLT BATTERY AS POWER SUPPLY.)

SNARE/BRUSH NOISE

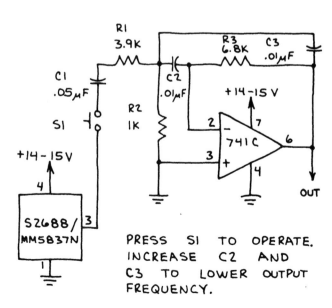

OUT

PRESS S1 TO OPERATE. INCREASE C2 AND C3 TO LOWER OUTPUT FREQUENCY.

39

NOTES

TTL/LS INTEGRATED CIRCUITS

INTRODUCTION

TTL IS THE BEST ESTABLISHED AND MOST DIVERSIFIED IC FAMILY. LS IS FUNCTIONALLY IDENTICAL TO TTL BUT IS SLIGHTLY FASTER AND USES 80% LESS POWER. TTL/LS CHIPS REQUIRE A REGULATED 4.75-5.25 VOLT POWER SUPPLY. HERE'S A SIMPLE BATTERY SUPPLY:

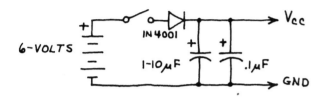

THE DIODE DROPS THE BATTERY VOLTAGE TO A SAFE LEVEL. BOTH CAPACITORS SHOULD BE INSTALLED ON THE TTL/LS CIRCUIT BOARD. CIRCUITS WITH LOTS OF TTL/LS CHIPS CAN USE LOTS OF CURRENT. USE A COMMERCIAL 5 VOLT LINE POWERED SUPPLY TO SAVE BATTERIES. OR MAKE YOUR OWN. (SEE THE 7805 ON PAGE 86.)

OPERATING REQUIREMENTS

1. V_{CC} MUST <u>NOT</u> EXCEED 5.25 VOLTS.

2. INPUT SIGNALS MUST NEVER EXCEED V_{CC} AND SHOULD NOT FALL BELOW GND.

3. UNCONNECTED TTL/LS INPUTS USUALLY ASSUME THE H STATE ... BUT <u>DON'T COUNT ON IT</u>! IF AN INPUT IS SUPPOSED TO BE FIXED AT H, CONNECT IT TO V_{CC}.

4. IF AN INPUT IS SUPPOSED TO BE FIXED AT L, CONNECT IT TO GND.

5. CONNECT UNUSED AND/NAND/OR INPUTS TO A USED INPUT OF THE SAME CHIP.

6. FORCE OUTPUTS OF UNUSED GATES H TO SAVE CURRENT (NAND—ONE INPUT H; NOR—ALL INPUTS L).

7. USE AT LEAST ONE DECOUPLING CAPACITOR (0.01—0.1 μF) FOR EVERY 5-10 GATE PACKAGES, ONE FOR EVERY 2-5 COUNTERS AND REGISTERS AND ONE FOR EACH ONE-SHOT. DECOUPLING CAPACITORS NEUTRALIZE THE HEFTY POWER SUPPLY SPIKES THAT OCCUR WHEN A TTL/LS OUTPUT CHANGES STATES. THEY MUST HAVE SHORT LEADS AND BE CONNECTED FROM V_{CC} TO GND AS NEAR THE TTL/LS ICs AS POSSIBLE.

8. AVOID LONG WIRES WITHIN CIRCUITS

9. IF THE POWER SUPPLY IS NOT ON THE CIRCUIT BOARD, CONNECT A 1-10μF CAPACITOR ACROSS THE POWER LEADS WHERE THEY ARRIVE AT THE BOARD.

INTERFACING TTL/LS

1. 1 TTL OUTPUT WILL DRIVE UP TO 10 TTL OR 20 LS INPUTS.

2. 1 LS OUTPUT WILL DRIVE UP TO 5 TTL OR 10 LS INPUTS.

3. TTL/LS LED DRIVERS:

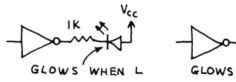

GLOWS WHEN L   GLOWS WHEN H

TTL/LS TROUBLESHOOTING

1. DO <u>ALL</u> INPUTS GO SOMEWHERE?

2. ARE <u>ALL</u> IC PINS INSERTED INTO THE BOARD OR SOCKET?

3. DOES THE CIRCUIT OBEY <u>ALL</u> TTL/LS OPERATING REQUIREMENTS?

4. HAVE YOU FORGOTTEN A CONNECTION?

5. HAVE YOU USED ENOUGH DECOUPLING CAPACITORS? ARE THEIR LEADS SHORT?

6. IS V_{CC} AT EACH CHIP WITHIN RANGE?

QUAD NAND GATE
7400/74LS00

THE BASIC BUILDING BLOCK CHIP
FOR THE ENTIRE TTL FAMILY. VERY
EASY TO USE. HUNDREDS OF APPLICATIONS.

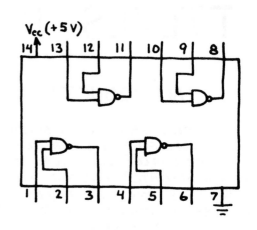

CONTROL GATE

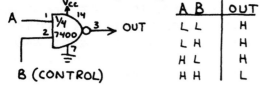

A	B	OUT
L	L	H
L	H	H
H	L	H
H	H	L

INVERTER

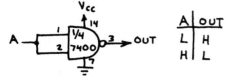

A	OUT
L	H
H	L

AND GATE

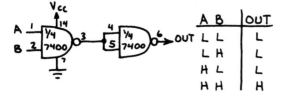

A	B	OUT
L	L	L
L	H	L
H	L	L
H	H	H

OR GATE

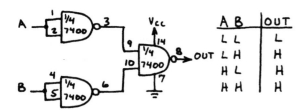

A	B	OUT
L	L	L
L	H	H
H	L	H
H	H	H

AND-OR GATE

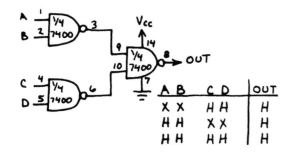

A	B	C	D	OUT
X	X	H	H	H
H	H	X	X	H
H	H	H	H	H

NOR GATE

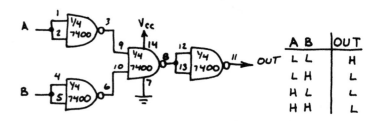

A	B	OUT
L	L	H
L	H	L
H	L	L
H	H	L

4-INPUT NAND GATE

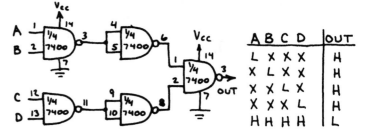

A	B	C	D	OUT
L	X	X	X	H
X	L	X	X	H
X	X	L	X	H
X	X	X	L	H
H	H	H	H	L

EXCLUSIVE-OR GATE

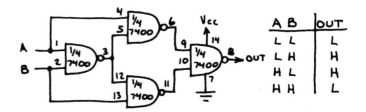

A	B	OUT
L	L	L
L	H	H
H	L	H
H	H	L

EXCLUSIVE-NOR GATE

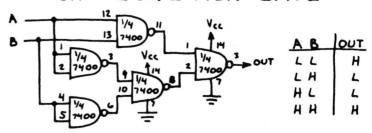

A	B	OUT
L	L	H
L	H	L
H	L	L
H	H	H

NOTE: PIN NUMBERS CAN BE
REARRANGED IF DESIRED.

QUAD NAND GATE (CONTINUED)
7400/74LS00

HALF ADDER

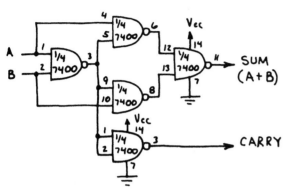

SUM (A+B)

CARRY

RS LATCH

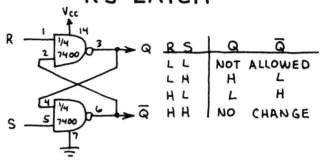

R	S	Q	$\overline{Q}$
L	L	NOT ALLOWED	
L	H	H	L
H	L	L	H
H	H	NO CHANGE	

D FLIP-FLOP

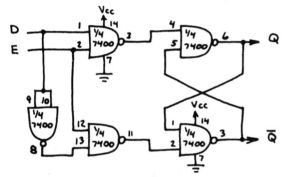

WHEN ENABLE (E) INPUT IS HIGH, Q OUTPUT FOLLOWS D INPUT. NO CHANGE WHEN E IS LOW.

GATED RS LATCH

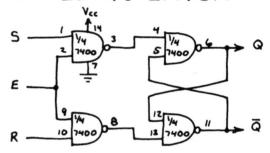

FUNCTIONS AS RS LATCH WHEN ENABLE (E) INPUT IS HIGH. IGNORES RS INPUTS WHEN E IS LOW.

LED DUAL FLASHER

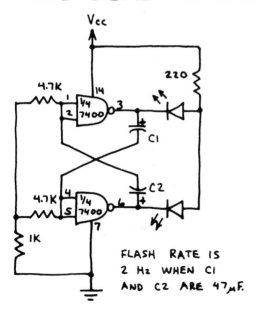

FLASH RATE IS 2 Hz WHEN C1 AND C2 ARE 47µF.

SWITCH DEBOUNCER

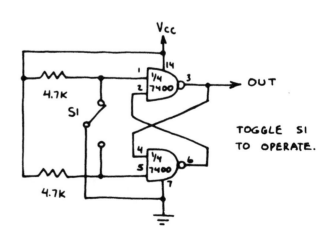

OUT

TOGGLE S1 TO OPERATE.

PROVIDES NOISE FREE OUTPUT FROM STANDARD SPDT TOGGLE SWITCH.

8-INPUT NAND GATE

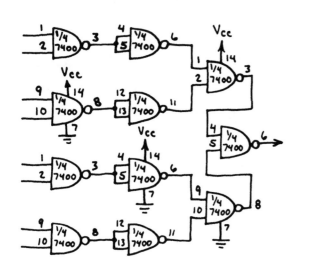

BCD DECODER

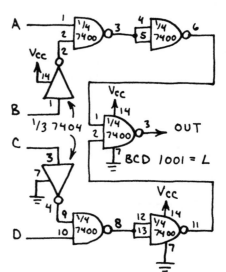

A	B	C	D	OUT
H	L	L	H	L
X	X	X	X	H

USE THIS METHOD TO DECODE ANY 4-BIT NIBBLE. JUST ADD OR REMOVE INPUT INVERTERS.

IC 1, 2 = 7400/74LS00

UNANIMOUS VOTE DETECTOR

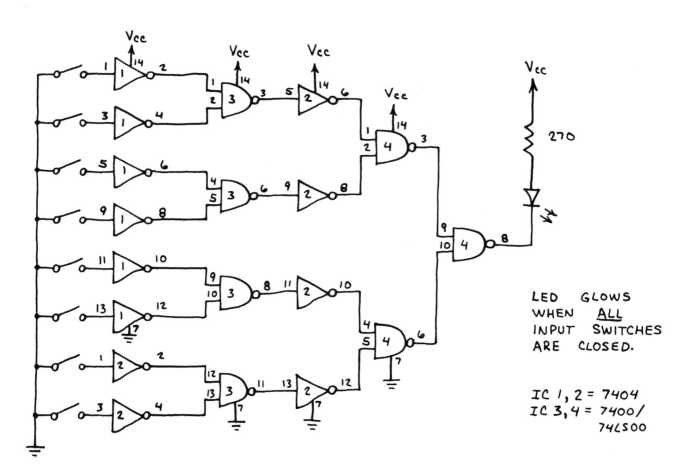

LED GLOWS WHEN <u>ALL</u> INPUT SWITCHES ARE CLOSED.

IC 1, 2 = 7404
IC 3, 4 = 7400/74LS00

QUAD AND GATE
7408/74LS08

ONE OF THE BASIC BUILDING BLOCK CHIPS. NOT AS VERSATILE, HOWEVER, AS THE 7400/74LS00 QUAD NAND GATE.

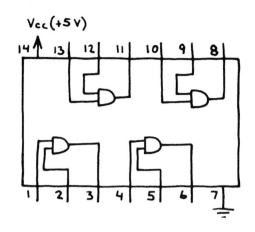

AND GATE BUFFER

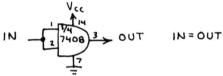

IN = OUT

USE FOR INTERFACING WITHOUT CHANGING LOGIC STATES.

DIGITAL TRANSMISSION GATE

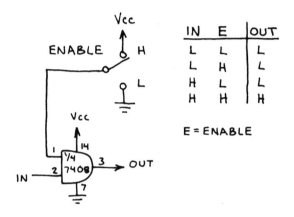

IN	E	OUT
L	L	L
L	H	L
H	L	L
H	H	H

E = ENABLE

NAND GATE

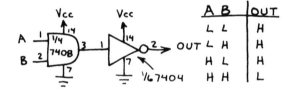

A	B	OUT
L	L	H
L	H	H
H	L	H
H	H	L

NOR GATE

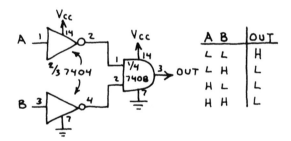

A	B	OUT
L	L	H
L	H	L
H	L	L
H	H	L

AND-OR-INVERT GATE

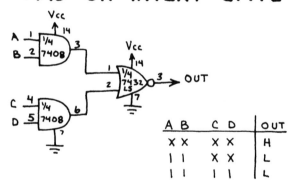

A B	C D	OUT
X X	X X	H
I I	X X	L
I I	I I	L

4-INPUT NAND GATE

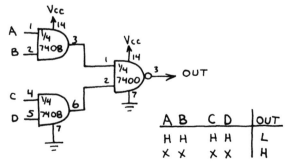

A B	C D	OUT
H H	H H	L
X X	X X	H

4-INPUT AND GATE

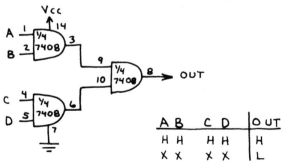

A B	C D	OUT
H H	H H	H
X X	X X	L

45

QUAD OR GATE
74LS32

FOUR 2-INPUT OR GATES.
NOT AS VERSATILE AS 7402/
74LS02 QUAD NOR GATE,
BUT VERY USEFUL IN SIMPLE
DATA SELECTORS.

Vcc (+5 v)

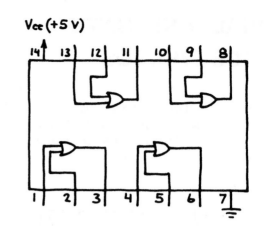

AND-OR CIRCUIT

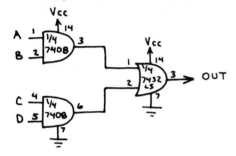

OUTPUT GOES HIGH WHEN BOTH
INPUTS OF EITHER OR BOTH AND
GATES ARE HIGH; OTHERWISE
THE OUTPUT IS LOW. THIS BASIC
CIRCUIT IS USED TO MAKE
DATA SELECTORS... AS SHOWN
BELOW ⌐→

NOR GATE

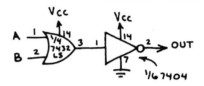

A B	OUT
L L	H
L H	L
H L	L
H H	L

NAND GATE

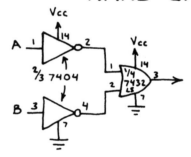

A B	OUT
L L	H
L H	H
H L	H
H H	L

2-INPUT DATA SELECTOR

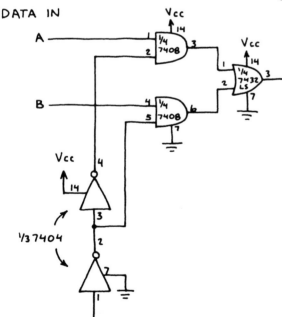

ADDRESS (DATA SELECT)

SELECTS 1-OF-2 INPUTS
AND TRANSMITS ITS
LOGIC STATE TO THE
OUTPUT.

ADDRESS	DATA	IN	OUT
A	B	A	
L	X	L	L
L	X	H	H
H	L	X	L
H	H	X	H

NOTE: FOR 3-INPUT DATA SELECTOR,
USE 74LS27 NOR GATE FOLLOWED
BY INVERTER AND PRECEEDED BY
74LS10 3-INPUT AND GATES.

QUAD NOR GATE
7402/74LS02

JUST AS VERSATILE AS THE
7400/74LS00 QUAD NAND GATE...
BUT NOT USED AS OFTEN.
ADD INVERTER (7404/74LS04)
TO BOTH INPUTS OF A NOR
GATE AND AN AND GATE IS
FORMED.

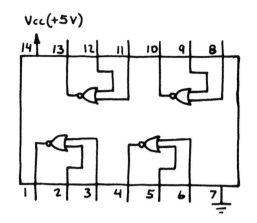

EXCLUSIVE-OR GATE

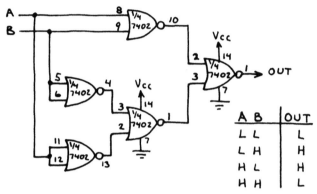

A B	OUT
L L	L
L H	H
H L	H
H H	L

THIS CIRCUIT IS EQUIVALENT
TO A BINARY HALF-ADDER.

ONE-SHOT

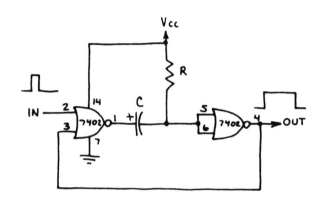

THIS CIRCUIT IS A MONOSTABLE
MULTIVIBRATOR OR PULSE STRETCHER.
AN INPUT PULSE TRIGGERS AN
OUTPUT PULSE WITH A DURATION
DETERMINED BY R AND C. OUTPUT
PULSE WIDTH IS APPROXIMATELY 0.8RC.

RS LATCH

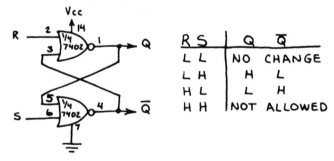

R S	Q	Q̄
L L	NO CHANGE	
L H	H	L
H L	L	H
H H	NOT ALLOWED	

AND GATE

A B	OUT
L L	L
L H	L
H L	L
H H	H

4-INPUT NOR GATE

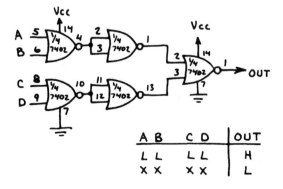

A B	C D	OUT
L L	L L	H
X X	X X	L

OR GATE

A B	OUT
L L	L
L H	H
H L	H
H H	H

DUAL 4-INPUT NAND GATE
74LS20

MANY DECODER AND ENCODER
APPLICATIONS. CAN BE USED
AS DUAL 3-INPUT NAND GATE
WITH ENABLE (CONTROL) INPUT
FOR EACH GATE.

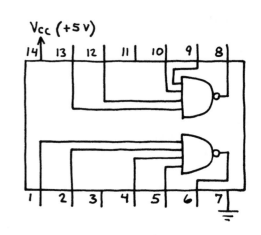

BCD DECODERS

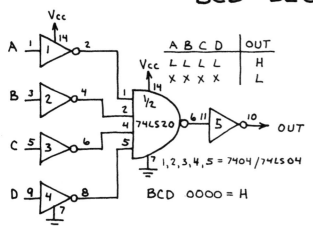

A	B	C	D	OUT
L	L	L	L	H
X	X	X	X	L

BCD 0000 = H

1,2,3,4,5 = 7404/74LS04

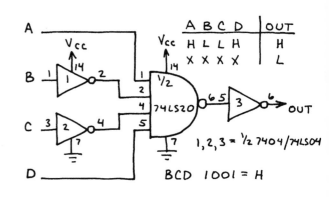

A	B	C	D	OUT
H	L	L	H	H
X	X	X	X	L

BCD 1001 = H

1,2,3 = ½ 7404/74LS04

OUTPUTS GO HIGH WHEN APPROPRIATE BCD WORD
APPEARS AT INPUTS DCBA. OUTPUTS STAY LOW
FOR ALL OTHER INPUTS. (OMIT FINAL INVERTER TO
PROVIDE ACTIVE LOW OUTPUT.) USE THIS METHOD TO
DECODE ANY 4-BIT NIBBLE.

DECIMAL-TO-BINARY CODED DECIMAL (BCD) ENCODER

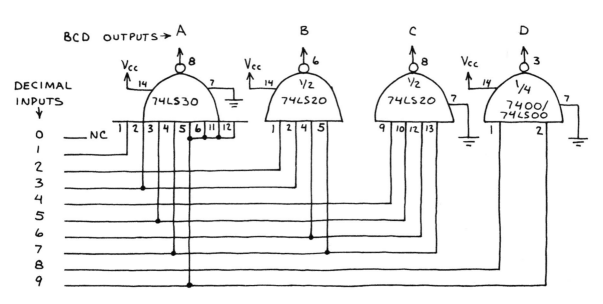

SELECTED INPUT SHOULD BE LOW AND ALL OTHER INPUTS SHOULD
BE HIGH. BCD EQUIVALENT WILL APPEAR AT THE OUTPUTS.

48

TRIPLE 3-INPUT NOR GATE
74LS27

USEFUL FOR DATA SELECTORS
AND NOR GATE FLIP-FLOPS
THAT REQUIRE CLEAR AND
PRESET INPUTS.

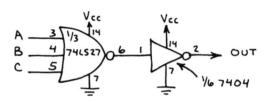

GATED RS LATCH

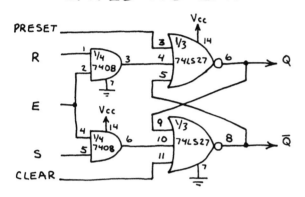

FUNCTIONS AS RS LATCH WHEN
E (ENABLE) INPUT IS HIGH. IGNORES
RS INPUTS WHEN E IS LOW.

3-INPUT OR GATE

3-INPUT DATA SELECTOR

SELECTS 1-OF-3 INPUTS AND TRANSMITS
ITS LOGIC STATE TO THE OUTPUT.

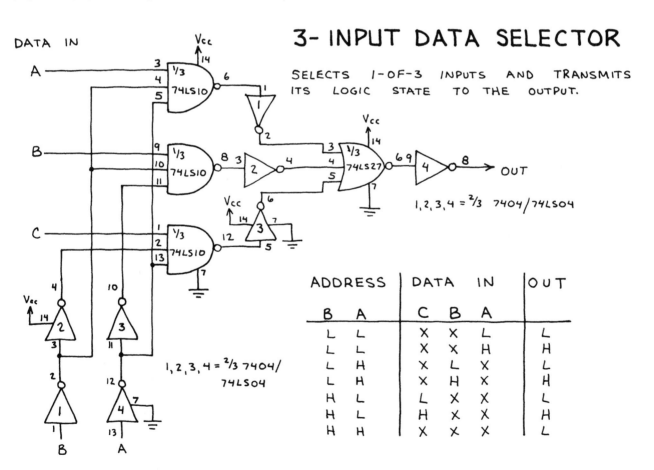

1,2,3,4 = ²/₃ 7404/74LS04

ADDRESS		DATA	IN		OUT
B	A	C	B	A	
L	L	X	X	L	L
L	L	X	X	H	H
L	H	X	L	X	L
L	H	X	H	X	H
H	L	L	X	X	L
H	L	H	X	X	H
H	H	X	X	X	L

1,2,3,4 = ²/₃ 7404/74LS04

ADDRESS (DATA SELECT)

49

8-INPUT NAND GATE
74LS30

HANDY FOR BYTE-SIZE (8-BIT) DECODING
APPLICATIONS. CAN DECODE UP TO
256 INPUT COMBINATIONS. ALSO
USEFUL AS PROGRAMMABLE NAND
GATE.

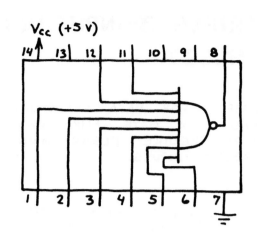

8-BIT DECODER

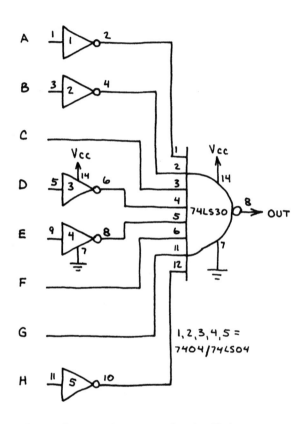

1,2,3,4,5 =
7404/74LS04

OUTPUT GOES LOW ONLY WHEN
INPUT IS LHHLLHLL (DECIMAL 100).
UP TO 256 INPUTS CAN BE DECODED
BY REARRANGING UP TO 8 INPUT
INVERTERS.

UNANIMOUS VOTE DETECTOR

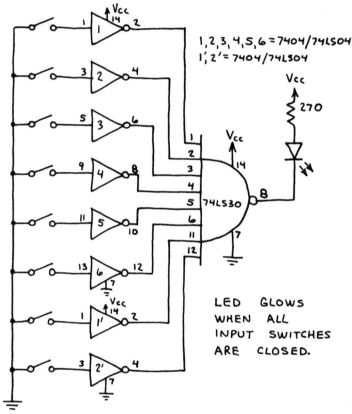

1,2,3,4,5,6 = 7404/74LS04
1',2' = 7404/74LS04

LED GLOWS
WHEN ALL
INPUT SWITCHES
ARE CLOSED.

PROGRAMMABLE NAND GATES

5-INPUT

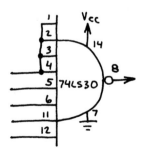

6-INPUT

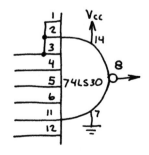

7-INPUT

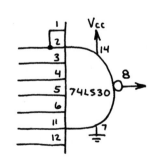

50

DUAL AND-OR-INVERT GATE
74LS51

VERY VERSATILE BUILDING BLOCK
CHIP. IDEAL FOR CUSTOMIZED
DATA SELECTORS, LATCHES
AND EXPANSION OF A SINGLE
INPUT TO AN AND-OR INPUT.

LATCH WITH ENABLE INPUT

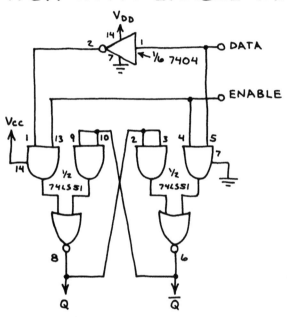

Q OUTPUT FOLLOWS DATA INPUT
WHEN ENABLE INPUT IS HIGH. NO
CHANGE WHEN ENABLE IS LOW.

TYPICAL AND-OR INPUT

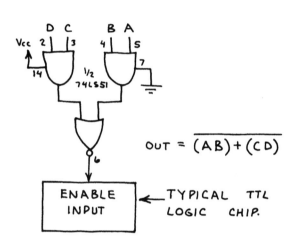

$$OUT = \overline{(AB) + (CD)}$$

THIS CIRCUIT SELECTS
1-OF-2 4-BIT WORDS.
NOTE THAT THE
SELECTED WORD IS
INVERTED AT THE
OUTPUTS. THE CIRCUIT
REQUIRES TWO
74LS51 CHIPS.

1-OF-2 DATA SELECTOR

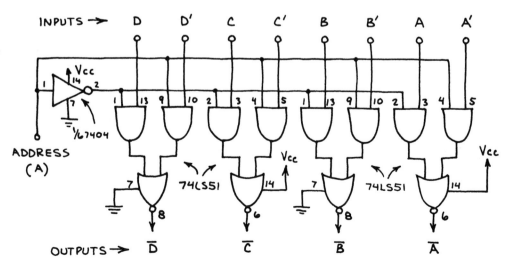

A	INPUT		OUT
	X	X	
H	X	L	H
H	X	H	L
L	L	X	H
L	H	X	L

DUAL NAND SCHMITT TRIGGER 74LS13

TWO 4-INPUT NAND GATES
WITH A SWITCHING THRESHOLD.
OUTPUTS GO LOW WHEN INPUTS
EXCEED 1.7 VOLTS. OUTPUTS GO
HIGH WHEN INPUTS FALL TO
0.9 VOLT. IF ANY INPUT IS LOW,
THE RESPECTIVE OUTPUT WILL
STAY HIGH AND THE GATE WILL
NOT TRIGGER.

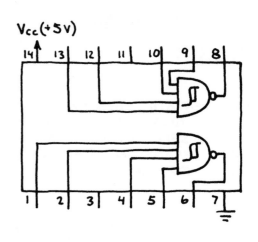

GATED THRESHOLD DETECTOR

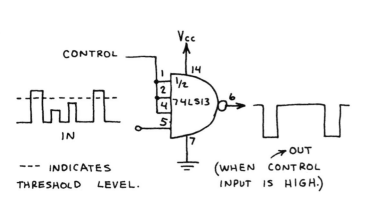

--- INDICATES
THRESHOLD LEVEL.

(WHEN CONTROL
INPUT IS HIGH.)

PHOTOTRANSISTOR RECEIVER

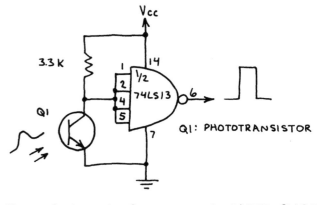

Q1: PHOTOTRANSISTOR

USE TO CLEAN UP INCOMING LIGHT PULSES.

GATED OSCILLATOR

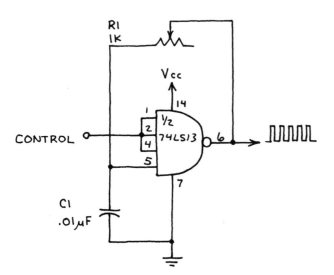

OSCILLATES WHEN CONTROL IS HIGH.
CHANGE R1 AND C1 TO CHANGE
FREQUENCY. OK TO USE THIS CIRCUIT
AS GATED CLOCK FOR LOGIC CIRCUITS.

TWO-STATE LED FLASHER

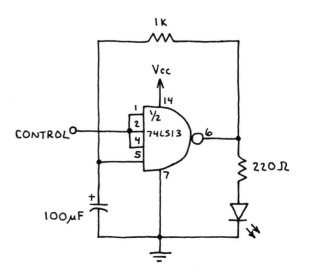

LED FLASHES TWICE EACH SECOND
WHEN CONTROL INPUT IS HIGH.
LED STAYS ON AND DOES NOT
FLASH WHEN CONTROL IS LOW.

52

HEX INVERTER
7404/74LS04

VERY IMPORTANT IN ALMOST ALL LOGIC CIRCUITS. CHANGES AN INPUT TO ITS COMPLEMENT (i.e. H→L AND L→H).

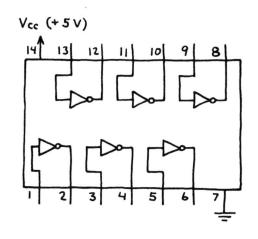

BOUNCEFREE SWITCH

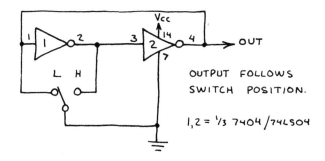

OUTPUT FOLLOWS SWITCH POSITION.

$1, 2 = \frac{1}{3}$ 7404/74LS04

AUDIO OSCILLATOR

OUTPUT TONE IS 4 KHz.

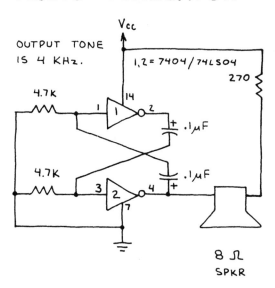

1,2 = 7404/74LS04

8 Ω SPKR

UNIVERSAL EXPANDER

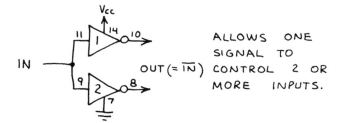

OUT(=$\overline{IN}$)

ALLOWS ONE SIGNAL TO CONTROL 2 OR MORE INPUTS.

1-OF-2 DEMULTIPLEXER

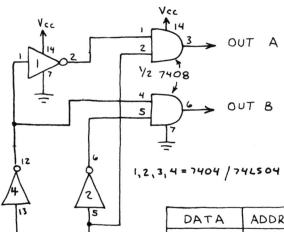

OUT A

OUT B

½ 7408

1,2,3,4 = 7404/74LS04

THIS CIRCUIT STEERS THE INPUT BIT TO THE OUTPUT SELECTED BY THE ADDRESS.

THIS TECHNIQUE CAN BE USED TO MAKE MULTIPLE OUTPUT DEMULTIPLEXERS.

DATA	ADDRESS	OUT A	OUT B
L	L	L	L
H	L	H	L
L	H	L	L
H	H	L	H

DATA

A (ADDRESS)

53

HEX 3-STATE BUS DRIVER
74LS367

EACH GATE FUNCTIONS AS A
NON-INVERTING BUFFER WHEN
ITS ENABLE INPUT (G1 OR G2)
IS LOW. OTHERWISE EACH GATE'S
OUTPUT ENTERS THE HIGH
IMPEDANCE (HI-Z) STATE.

HERE'S THE
TRUTH TABLE:

G	IN	OUT
H	X	HI-Z
L	L	L
L	H	H

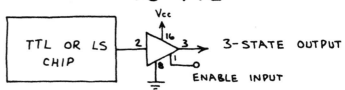

1-OF-2 DATA SELECTOR

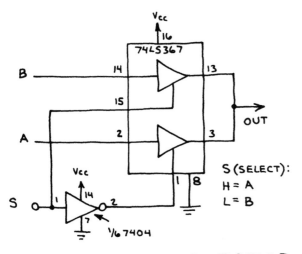

S (SELECT):
H = A
L = B

1-OF-2 DATA SELECTOR

INPUT
WORDS

SELECTS
1-OF-2
2-BIT WORDS.

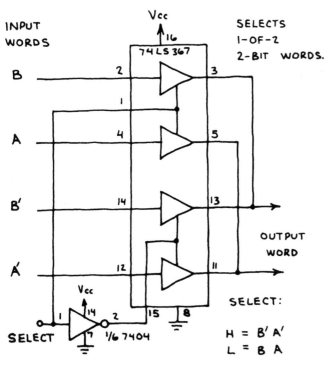

OUTPUT
WORD

SELECT:
H = B' A'
L = B A

ADDING 3-STATE OUTPUT TO TTL

TTL OR LS
CHIP

3-STATE OUTPUT

ENABLE INPUT

BIDIRECTIONAL DATA BUS

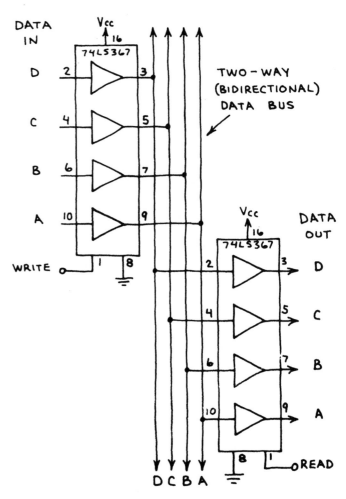

TWO-WAY
(BIDIRECTIONAL)
DATA BUS

DATA
IN

DATA
OUT

54

HEX 3-STATE BUS DRIVER 74LS368

EACH GATE FUNCTIONS AS AN INVERTER WHEN ITS ENABLE INPUT (G1 OR G2) IS LOW. OTHERWISE EACH GATE'S OUTPUT ENTERS THE HIGH IMPEDANCE (HI-Z) STATE.

HERE'S THE TRUTH TABLE:

G	IN	OUT
H	X	HI-Z
L	L	H
L	H	L

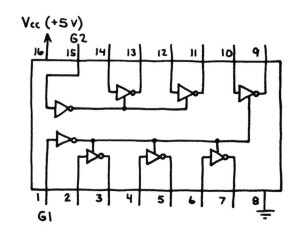

GATED LED FLASHER

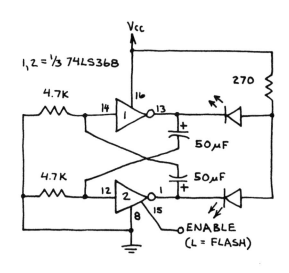

1,2 = 1/3 74LS368

BIDIRECTIONAL DATA BUS

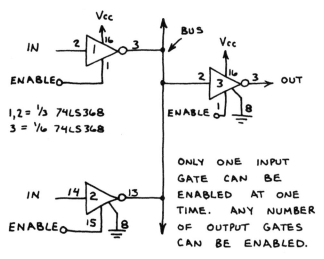

1,2 = 1/3 74LS368
3 = 1/6 74LS368

ONLY ONE INPUT GATE CAN BE ENABLED AT ONE TIME. ANY NUMBER OF OUTPUT GATES CAN BE ENABLED.

GATED TONE SOURCE

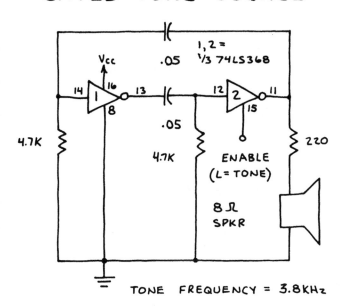

1,2 = 1/3 74LS368

TONE FREQUENCY = 3.8KHz

BOUNCELESS SWITCH (WITH ENABLE)

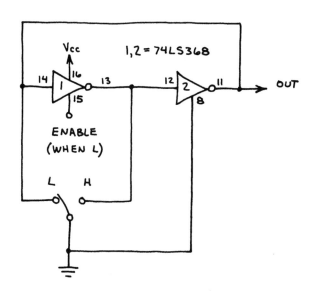

1,2 = 74LS368

4-BIT MAGNITUDE COMPARATOR 74LS85

COMPARES TWO 4-BIT WORDS. INDICATES WHICH IS LARGER OR IF THEY ARE EQUAL.

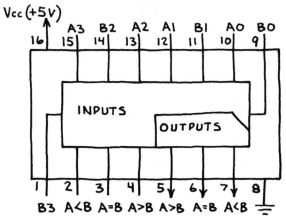

8-BIT COMPARATOR

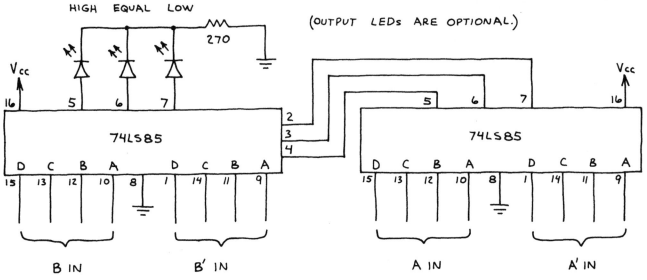

(OUTPUT LEDs ARE OPTIONAL.)

BINARY HI-LO GAME

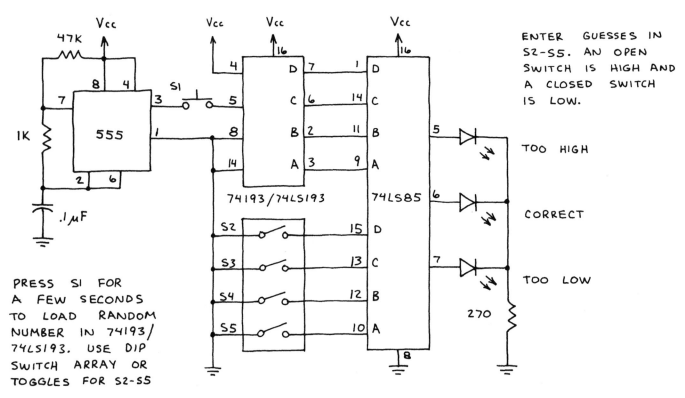

ENTER GUESSES IN S2-S5. AN OPEN SWITCH IS HIGH AND A CLOSED SWITCH IS LOW.

TOO HIGH

CORRECT

TOO LOW

PRESS S1 FOR A FEW SECONDS TO LOAD RANDOM NUMBER IN 74193/74LS193. USE DIP SWITCH ARRAY OR TOGGLES FOR S2-S5

56

BCD-TO-DECIMAL DECODER
7441

DECODES 4-BIT BCD INPUT INTO
1-OF-10 OUTPUTS. SELECTED
OUTPUT GOES LOW; ALL OTHERS
STAY HIGH. ORIGINALLY DESIGNED
TO DRIVE GASEOUS GLOW DISCHARGE
TUBES. ALL OUTPUTS GO HIGH FOR
BINARY INPUTS EXCEEDING HLLH (1001).

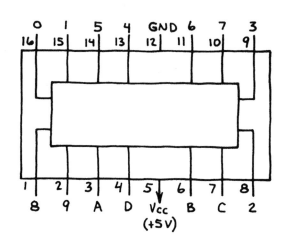

1-OF-10 DECODED COUNTER

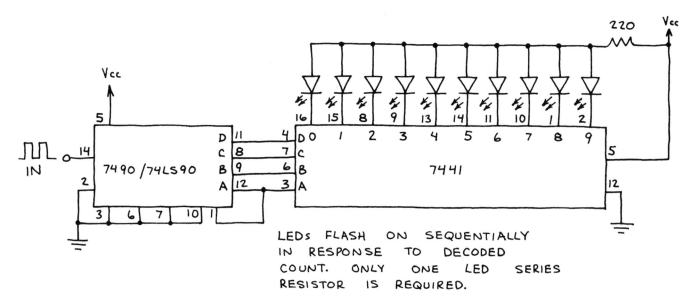

LEDs FLASH ON SEQUENTIALLY
IN RESPONSE TO DECODED
COUNT. ONLY ONE LED SERIES
RESISTOR IS REQUIRED.

10-NOTE TONE SEQUENCER

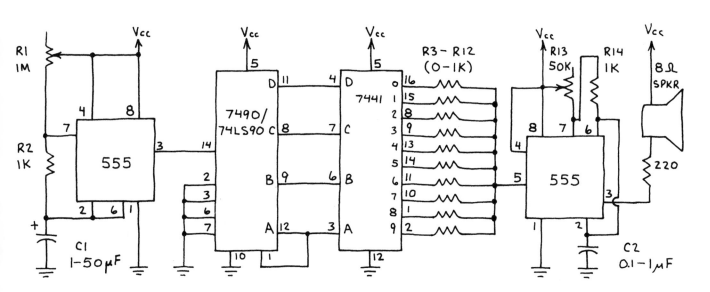

INCREASE C1 TO DECREASE TEMPO. INCREASE C2 TO INCREASE TONE
FREQUENCIES. TONES ARE DETERMINED BY R3-R12.

57

BCD-TO-7 SEGMENT DECODER/DRIVER

7447 / 74LS47

CONVERTS BCD DATA INTO FORMAT SUITABLE FOR PRODUCING DECIMAL DIGITS ON COMMON ANODE LED 7-SEGMENT DISPLAY. WHEN LAMP TEST INPUT IS LOW, ALL OUTPUTS ARE LOW (ON). WHEN BI/RBO (BLANKING INPUT) IS LOW, ALL OUTPUTS ARE HIGH (OFF). WHEN DCBA INPUT IS LLLL (DECIMAL 0) AND RBI (RIPPLE BLANKING INPUT) IS LOW, ALL OUTPUTS ARE HIGH (OFF). THIS PERMITS UNWANTED LEADING 0's IN A ROW OF DIGITS TO BE BLANKED.

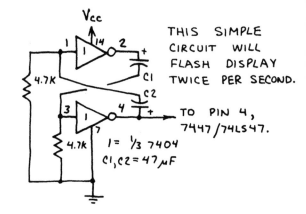

MANUALLY SWITCHED DISPLAY

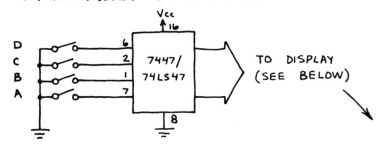

TO DISPLAY (SEE BELOW)

DISPLAY FLASHER

THIS SIMPLE CIRCUIT WILL FLASH DISPLAY TWICE PER SECOND.

TO PIN 4, 7447/74LS47.

$1 = \frac{1}{3}$ 7404
$C1, C2 = 47 \mu F$

0-9 SECOND/MINUTE TIMER

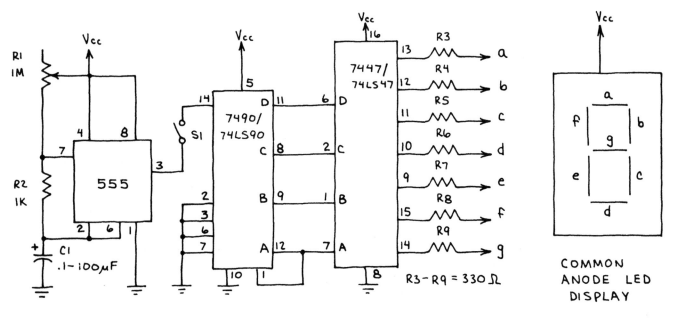

R3-R9 = 330 Ω

COMMON ANODE LED DISPLAY

CLOSE S1 TO START TIMING CYCLE. CALIBRATE 555 FOR 1 PULSE (COUNT) PER SECOND OR 1 COUNT PER MINUTE BY ADJUSTING R1.

58

BCD-TO-7-SEGMENT DECODER/DRIVER 7448

CONVERTS BCD DATA INTO FORMAT SUITABLE FOR PRODUCING DECIMAL DIGITS ON COMMON CATHODE LED 7-SEGMENT DISPLAY.

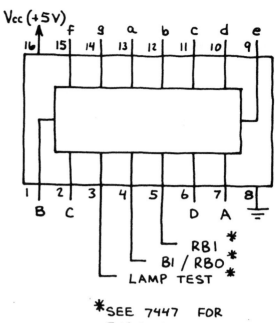

RBI *
BI / RBO **
LAMP TEST

* SEE 7447 FOR EXPLANATIONS.

DISPLAY DIMMER

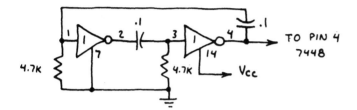

TO PIN 4 7448

0-99 TWO DIGIT COUNTER

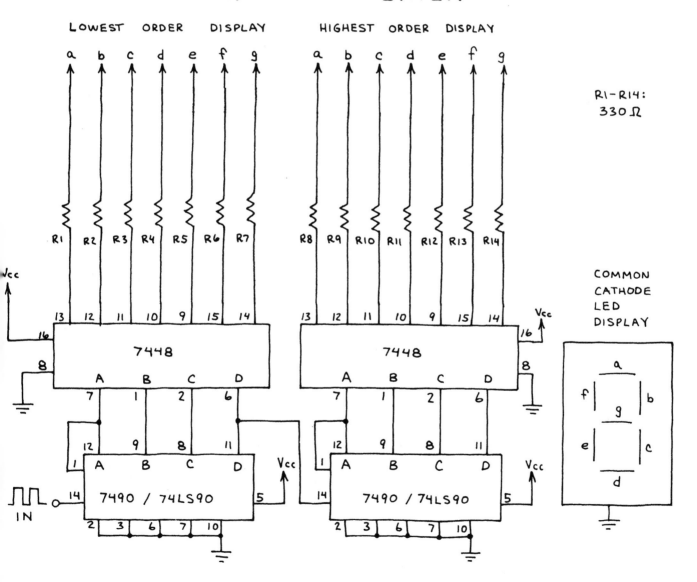

LOWEST ORDER DISPLAY

HIGHEST ORDER DISPLAY

R1-R14: 330 Ω

COMMON CATHODE LED DISPLAY

3-LINE TO 8-LINE DECODER
74LS138

EACH 3-BIT ADDRESS DRIVES
ONE OUTPUT LOW. ALL
OTHERS STAY HIGH. THIS
CHIP HAS THREE ENABLE
INPUTS. WHEN E2 IS HIGH,
ALL OUTPUTS ARE HIGH. WHEN
E1 IS LOW, ALL OUTPUTS
ARE HIGH. TO ENABLE CHIP,
MAKE E1 HIGH AND E2 LOW.
(NOTE: E2 = E2A + E2B.)

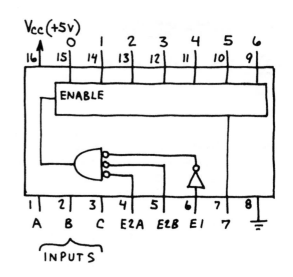

I-TO-8 DEMULTIPLEXER

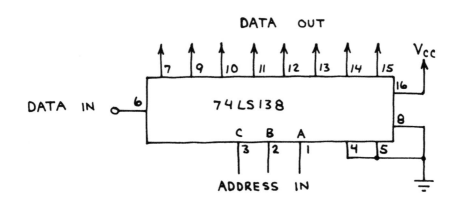

INPUT DATA (H
OR L) IS PASSED
TO SELECTED
OUTPUT.

2-TO-8 STEP SEQUENCER

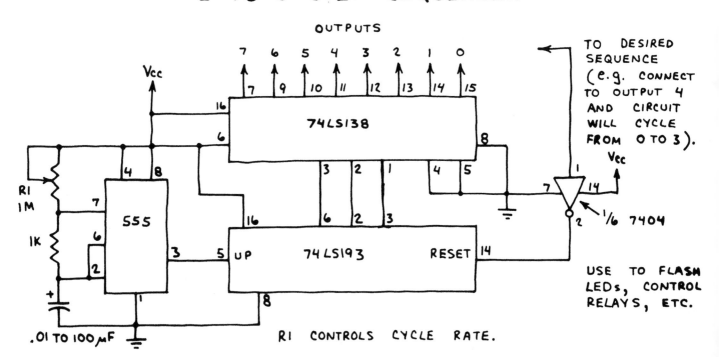

TO DESIRED
SEQUENCE
(e.g. CONNECT
TO OUTPUT 4
AND CIRCUIT
WILL CYCLE
FROM 0 TO 3).

1/6 7404

USE TO FLASH
LEDs, CONTROL
RELAYS, ETC.

R1 CONTROLS CYCLE RATE.

4-LINE TO 16-LINE DECODER 74154

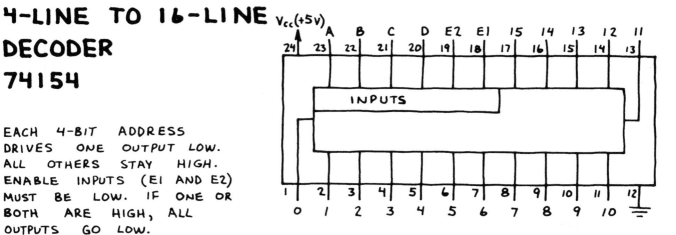

EACH 4-BIT ADDRESS DRIVES ONE OUTPUT LOW. ALL OTHERS STAY HIGH. ENABLE INPUTS (E1 AND E2) MUST BE LOW. IF ONE OR BOTH ARE HIGH, ALL OUTPUTS GO LOW.

1-TO-16 DEMULTIPLEXER

SELECTED OUTPUT IS LOW WHEN DATA IN IS LOW. IF DATA IN IS HIGH, SELECTED OUTPUT IS HIGH.

BACK AND FORTH FLASHER

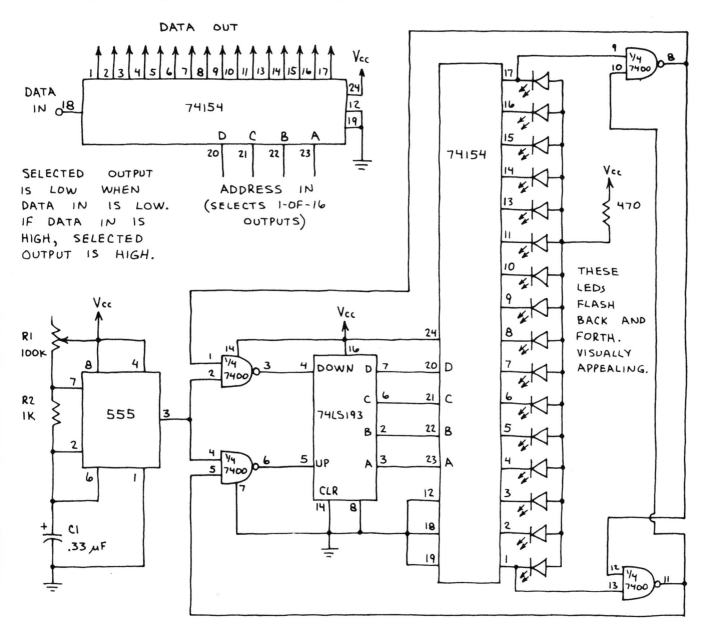

THESE LEDS FLASH BACK AND FORTH. VISUALLY APPEALING.

INCREASE R1 TO SLOW FLASH RATE.

QUAD 1-OF-2 DATA SELECTOR
74LS157

FOUR 2-LINE TO 1-LINE MULTIPLEXERS.
MANY USES IN ROUTING DATA. ALL
4 DATA SELECTORS ARE ENABLED
WHEN PIN 15 IS LOW.

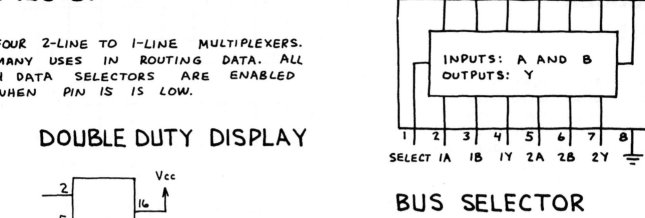

DOUBLE DUTY DISPLAY

BUS A

BUS B

74LS157

7-SEGMENT
DECODER
(7447/7448)
AND
7-SEGMENT
DISPLAY.

SELECT

LOW = BUS A
HIGH = BUS B

BUS SELECTOR

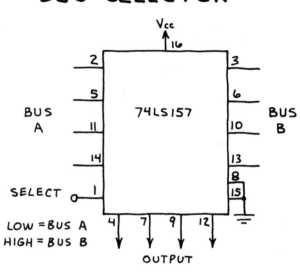

BUS A

BUS B

74LS157

SELECT

LOW = BUS A
HIGH = BUS B

OUTPUT

WORD SORTER

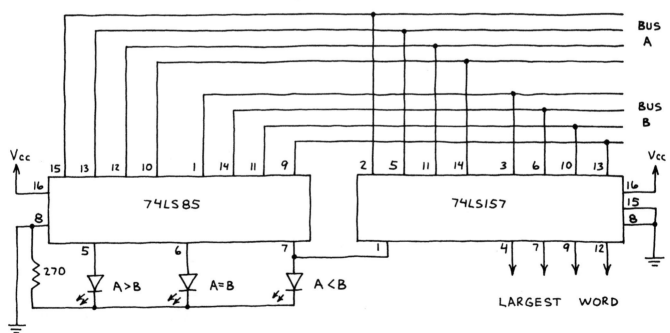

BUS A

BUS B

74LS85

74LS157

270

A>B

A=B

A<B

LARGEST WORD

THIS CIRCUIT CONTINUALLY MONITORS TWO DATA BUSES. BUS WITH
HIGHEST MAGNITUDE DATA WORD IS ROUTED AUTOMATICALLY TO OUTPUT.

1-OF-8 DATA SELECTOR
74LS151

EQUIVALENT TO 8-LINE TO 1-LINE MULTIPLEXER.

PROGRAMMABLE GATE

3-BIT ADDRESS SELECTS ONE
SWITCH AND APPLIES ITS STATUS
(OPEN = HIGH AND CLOSED = LOW) TO
THE OUTPUT. ANY 3-INPUT
LOGIC FUNCTION CAN BE
PROGRAMMED IN SECONDS.

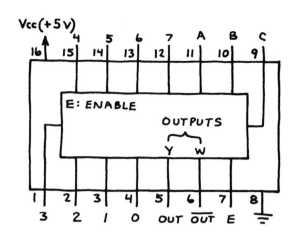

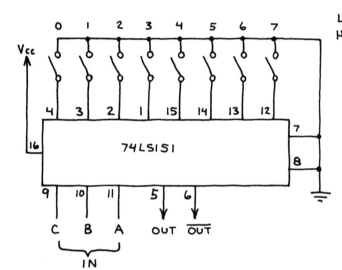

PATTERN GENERATOR

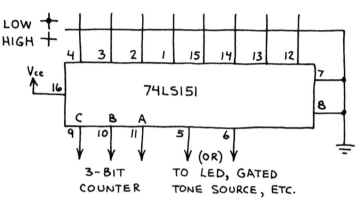

PROGRAM ANY DESIRED LOW-HIGH BIT
PATTERN. THEN PLAY IT BACK.

OCTAL KEYBOARD ENCODER

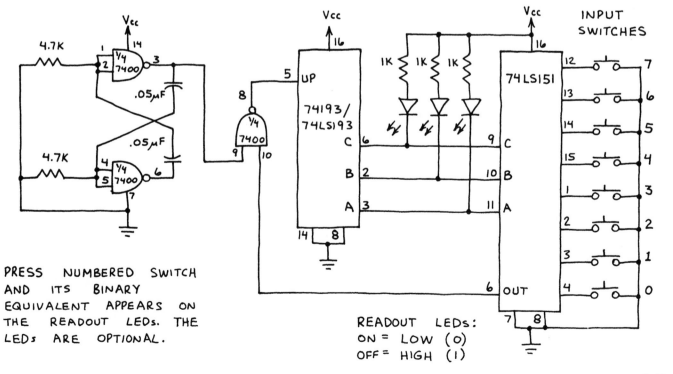

PRESS NUMBERED SWITCH
AND ITS BINARY
EQUIVALENT APPEARS ON
THE READOUT LEDs. THE
LEDs ARE OPTIONAL.

READOUT LEDs:
ON = LOW (0)
OFF = HIGH (1)

DUAL ONE-SHOT
74LS123

TWO FULLY INDEPENDENT
MONOSTABLE MULTIVIBRATORS.
BOTH ARE RETRIGGERABLE.
PINS DESIGNATED R AND R/C
ARE FOR EXTERNAL TIMING
RESISTOR AND CAPACITOR.

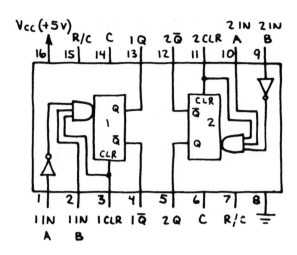

BASIC ONE-SHOT

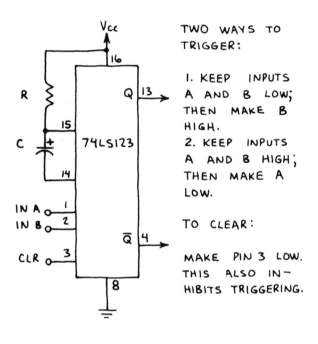

TWO WAYS TO
TRIGGER:

1. KEEP INPUTS
A AND B LOW;
THEN MAKE B
HIGH.
2. KEEP INPUTS
A AND B HIGH;
THEN MAKE A
LOW.

TO CLEAR:

MAKE PIN 3 LOW.
THIS ALSO IN-
HIBITS TRIGGERING.

MISSING PULSE DETECTOR

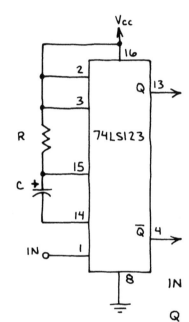

Q OUTPUT STAYS
HIGH SO LONG AS
INCOMING PULSES
ARRIVE BEFORE ONE-
SHOT TIMING PERIOD
RUNS OUT.

ADJUST R AND C
TO GIVE TIMING
PERIOD ABOUT 1/3
LONGER THAN THE
INTERVAL BETWEEN
INCOMING PULSES.

OPERATION:

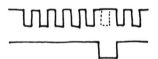

TONE STEPPER

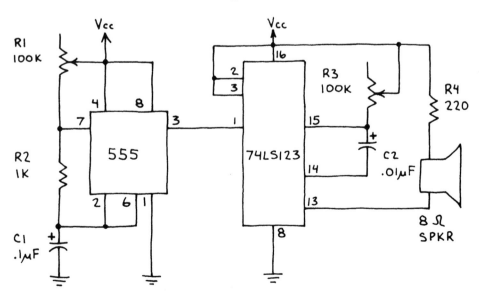

THIS CIRCUIT STEPS
ACROSS A RANGE
OF TONES WHEN R1
AND/OR R3 ARE
ADJUSTED. VERY
UNUSUAL SOUND
EFFECTS.

CHANGE C1 AND C2
FOR OTHER TONE
RANGES. ALSO, TRY
PHOTORESISTORS FOR
R1 AND R3.

64

DUAL D FLIP-FLOP
7474/74LS74

TWO D (DATA) FLIP-FLOPS IN A
SINGLE PACKAGE. DATA AT D
INPUT IS STORED AND MADE
AVAILABLE AT Q OUTPUT WHEN
CLOCK PULSE (Φ) GOES HIGH.
HERE'S THE TRUTH TABLE:

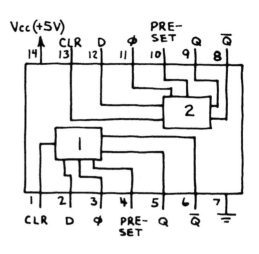

PRESET	CLEAR	CLOCK	D	Q	Q̄
L	H	X	X	H	L
H	L	X	X	L	H
H	H	↑	H	H	L
H	H	↑	L	L	H

Φ IS CLOCK INPUT.

↑ IS RISING EDGE OF CLOCK
PULSE.

2-BIT STORAGE REGISTER

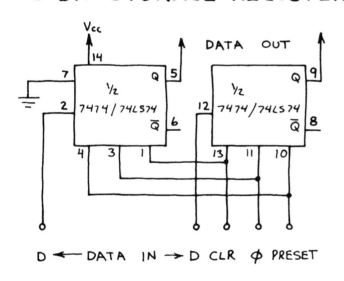

D ← DATA IN → D   CLR   Φ   PRESET

PHASE DETECTOR

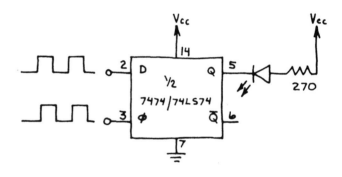

THE LED GLOWS WHEN INPUT
FREQUENCIES F1 AND F2 ARE
UNEQUAL OR OUT OF PHASE.
F1 AND F2 SHOULD BE
SQUARE WAVES.

WAVE SHAPER

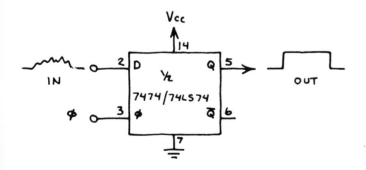

DIVIDE-BY-TWO COUNTER

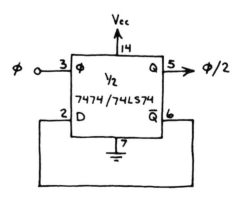

DUAL J-K FLIP-FLOP 7473

TWO JK FLIP-FLOPS IN A SINGLE PACKAGE. NOTE THE CLEAR INPUTS. THESE FLIP-FLOPS WILL TOGGLE (SWITCH OUTPUT STATES) IN RESPONSE TO INCOMING CLOCK PULSES WHEN BOTH J ANK J INPUTS ARE HIGH. HERE'S THE TRUTH TABLE:

CLEAR	CLOCK	J	K	Q	Q̄
L	X	X	X	L	H
H	⎍	H	L	H	L
H	⎍	L	H	L	H
H	⎍	H	H	TOGGLE	

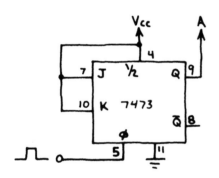

J Q̄ Q GND K Q Q̄
14 13 12 11 10 9 8

1   2

1 2 3 4 5 6 7
Φ CLR K   Φ CLR J

Vcc (+5V)

Φ IS CLOCK INPUT.

DIVIDE-BY-TWO

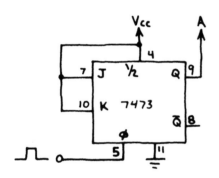

BINARY COUNTERS

THE THREE CIRCUITS ON THIS PAGE ARE BINARY COUNTERS THAT COUNT UP TO THE MAXIMUM COUNT AND AUTOMATICALLY RECYCLE. CONNECT A DECODER TO OUTPUT OF DIVIDE-BY-THREE AND DIVIDE-BY-FOUR COUNTERS TO OBTAIN ONE-OF-THREE AND ONE-OF-FOUR OPERATION. THIS TRUTH TABLE SUMMARIZES OPERATION OF THESE COUNTERS:

DIVIDE-BY:	TWO	THREE		FOUR	
OUTPUTS:	A	B	A	B	A
	L	L	L	L	L
	H	L	H	L	H
		H	L	H	L
				H	H

DIVIDE-BY-THREE

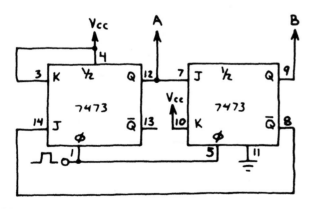

DIVIDE-BY-FOUR

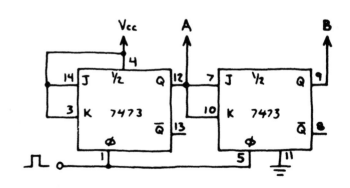

66

DUAL J-K FLIP-FLOP 7476

TWO JK FLIP-FLOPS IN A
SINGLE PACKAGE. SIMILAR
TO 7473/74LS73 BUT HAS
BOTH PRESET AND CLEAR
INPUTS. FLIP-FLOPS WILL
TOGGLE (SWITCH OUTPUT
STATES) IN RESPONSE TO
INCOMING CLOCK PULSES WHEN
BOTH J AND K INPUTS ARE
HIGH. HERE'S THE TRUTH TABLE:

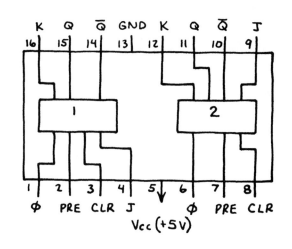

PRE = PRESET
CLR = CLEAR
ϕ = CLOCK (OR CLK)

TOGGLE = FLIP-FLOP SWITCHES
OUTPUT STATES IN
RESPONSE TO CLOCK
PULSES.

PRE	CLR	CLK	J	K	Q	$\bar{Q}$
L	H	X	X	X	H	L
H	L	X	X	X	L	H
H	H	⊓	H	L	H	L
H	H	⊓	L	H	L	H
H	H	⊓	H	H	TOGGLE	

4-BIT SERIAL SHIFT REGISTER

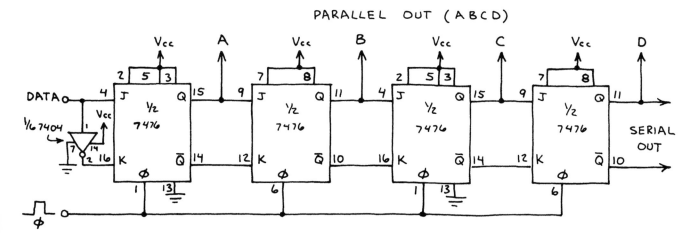

4-BIT BINARY UP COUNTER

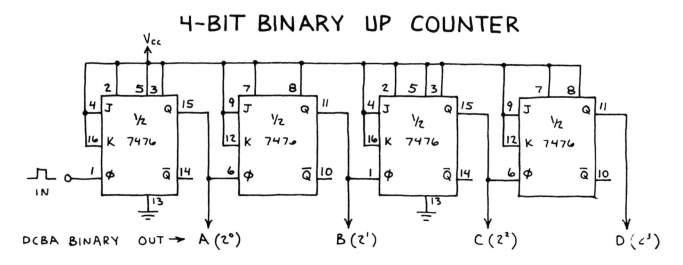

67

QUAD LATCH
7475/74LS75

A 4-BIT BISTABLE LATCH. PRIMARILY USED TO STORE THE COUNT IN DECIMAL COUNTING UNITS. NOTE THAT BOTH Q AND $\bar{Q}$ OUTPUTS ARE PROVIDED. ALSO NOTE THE E (ENABLE) INPUTS. WHEN E IS HIGH, Q FOLLOWS D.

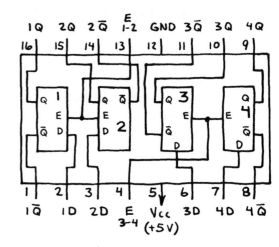

4-BIT DATA LATCH

DATA ON BUS APPEARS AT OUTPUTS WHEN LATCH INPUT IS HIGH. DATA ON BUS WHEN LATCH INPUT GOES LOW IS STORED UNTIL LATCH INPUT GOES HIGH. (LATCH INPUT CONTROLS BOTH ENABLE INPUTS.) TWO QUAD LATCHES CAN BE USED AS AN 8-BIT DATA LATCH.

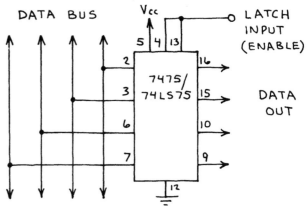

DECIMAL COUNTING UNIT

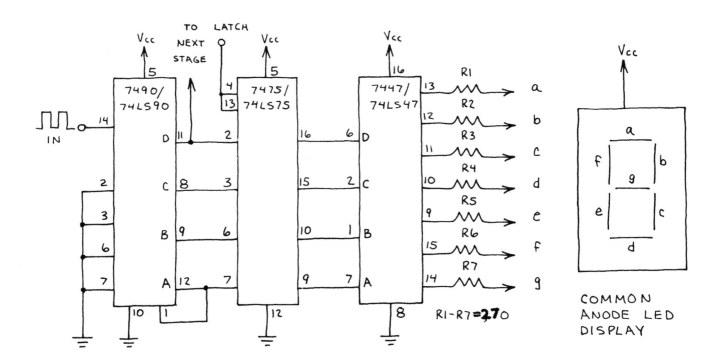

EXPANDABLE DECADE COUNTER. FOR TWO DIGIT COUNT, CONNECT PIN 11 OF 7490/74LS90 OF FIRST UNIT TO INPUT OF SECOND UNIT. A LOW AT THE LATCH INPUT FREEZES THE DATA BEING DISPLAYED.

QUAD D FLIP-FLOP
74LS175

HANDY PACKAGE OF FOUR D-TYPE
FLIP-FLOPS. DATA AT D-INPUTS
IS LOADED WHEN CLOCK GOES
HIGH. MAKING CLEAR INPUT
LOW MAKES ALL Q OUTPUTS LOW
AND $\bar{Q}$ OUTPUTS HIGH.

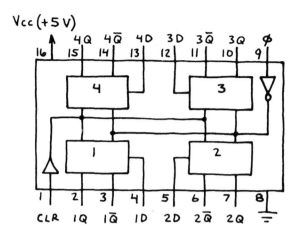

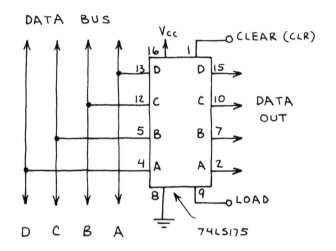

4-BIT DATA REGISTER

DATA ON BUS IS LOADED INTO
74LS175 WHEN LOAD INPUT
GOES HIGH. DATA IS THEN
STORED AND MADE AVAILABLE
AT OUTPUTS UNTIL NEW LOAD
PULSE ARRIVES.

MODULO-8 COUNTER

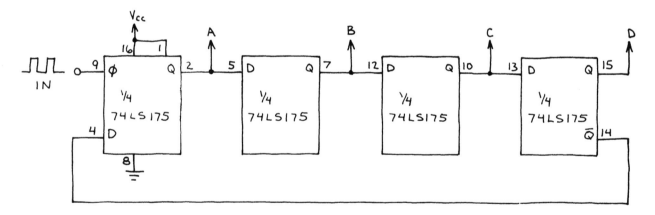

SERIAL IN/OUT, PARALLEL OUT SHIFT REGISTER

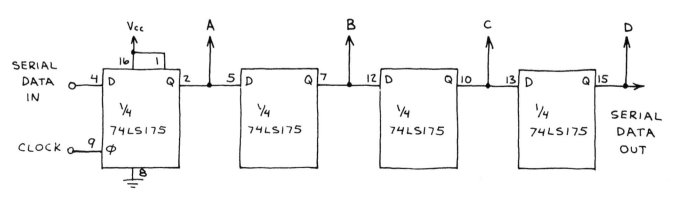

BCD (DECADE) COUNTER
7490/74LS90

ONE OF THE MOST POPULAR DECADE COUNTERS. EASILY USED FOR DIVIDE-BY-N COUNTERS. LESS EXPENSIVE THAN MORE SOPHISTICATED COUNTERS. RST INDICATES RESET PINS. THIS CHIP IS USUALLY USED IN DECIMAL COUNTING UNITS, BUT CIRCUITS ON THIS PAGE SHOW MANY OTHER POSSIBILITIES.

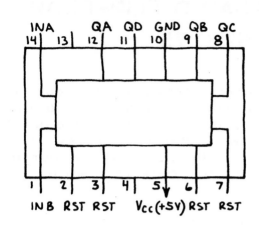

DIVIDE-BY-5 COUNTER

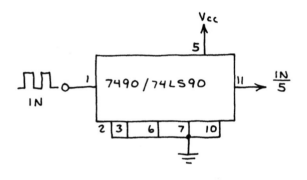

DIVIDE-BY-8 COUNTER

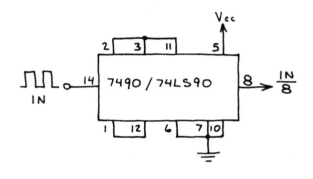

DIVIDE-BY-6 COUNTER

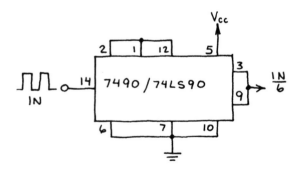

DIVIDE-BY-9 COUNTER

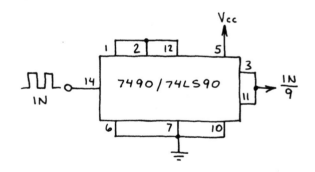

DIVIDE-BY-7 COUNTER

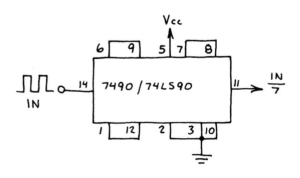

DIVIDE-BY-10 COUNTER

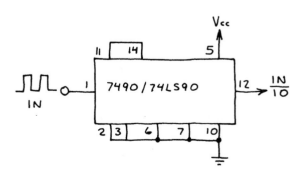

BCD (DECADE) COUNTER 74LS196

MORE SOPHISTICATED VERSION OF THE POPULAR 7490/74LS90 BCD COUNTER. INCLUDES 4-PRESET INPUTS WHICH PERMIT ANY BCD NUMBER TO BE LOADED WHEN PIN 1 IS MADE LOW. THE COUNTER IS CLEARED TO LLLL WHEN PIN 13 IS MADE LOW. ϕ INDICATES CLOCK INPUT.

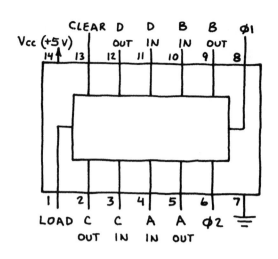

DECADE COUNTER

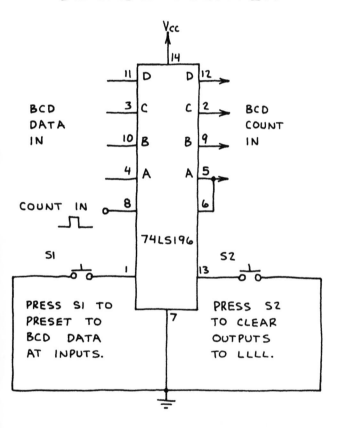

BCD DATA IN

BCD COUNT IN

COUNT IN

S1

S2

PRESS S1 TO PRESET TO BCD DATA AT INPUTS.

PRESS S2 TO CLEAR OUTPUTS TO LLLL.

4-BIT LATCH

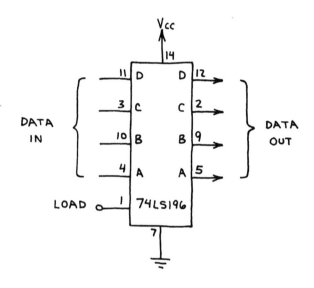

DATA IN

DATA OUT

LOAD

WHEN LOAD INPUT IS LOW, OUTPUTS FOLLOW INPUTS. NO CHANGE WHEN LOAD INPUT IS HIGH. NOTE THAT A PAIR OF 74LS196's CAN BE USED IN A DECIMAL COUNTING UNIT (COUNTER PLUS REGISTER).

DIVIDE-BY-5 COUNTER

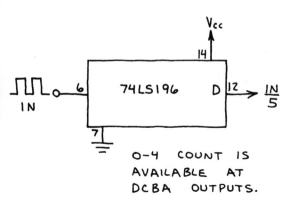

IN

$\frac{IN}{5}$

0-4 COUNT IS AVAILABLE AT DCBA OUTPUTS.

DIVIDE-BY-10 COUNTER

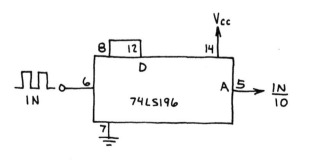

IN

$\frac{IN}{10}$

DIVIDE-BY-12 BINARY COUNTER
7492

OFTEN USED TO DIVIDE CONDITIONED
60 Hz PULSES FROM AC POWER
LINE INTO 10 Hz PULSES. OTHER
DIVIDER APPLICATIONS ALSO. RST
INDICATES RESET PINS.

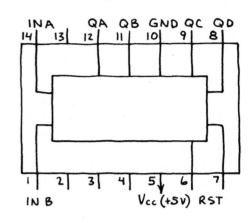

DIVIDE-BY-7 COUNTER

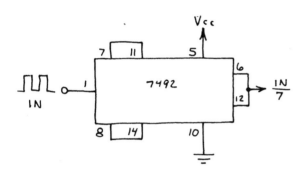

DIVIDE-BY-12 COUNTER

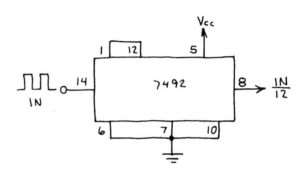

DIVIDE-BY-9 COUNTER

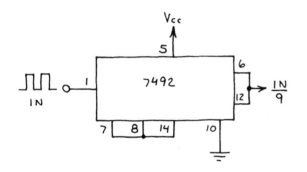

10-HZ PULSE SOURCE

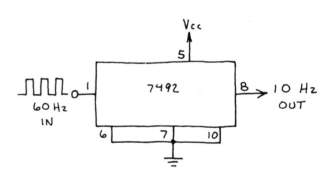

DIVIDE-BY-120 COUNTER

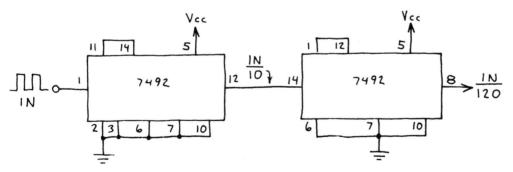

THIS METHOD OF
CASCADING COUNTERS
CAN BE USED TO
CREATE ANY
DIVIDE-BY-N
COUNTER.

72

4-BIT (BINARY) COUNTER
7493/74LS93

EASY TO USE 4-BIT BINARY
COUNTER. LESS EXPENSIVE
THAN MORE SOPHISTICATED
COUNTERS. RST INDICATES
RESET PINS. NOTE UNUSUAL
LOCATION OF POWER SUPPLY
PINS.

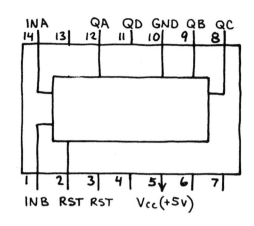

DIVIDE-BY-10 COUNTER

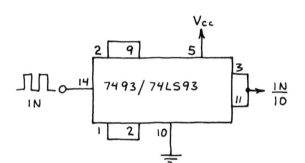

DIVIDE-BY-12 COUNTER

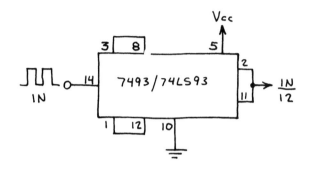

DIVIDE-BY-11 COUNTER

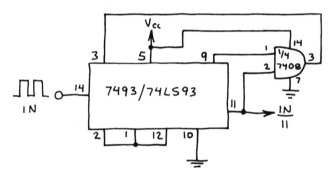

DIVIDE-BY-16 COUNTER

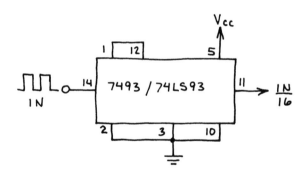

4-BIT BINARY COUNTER

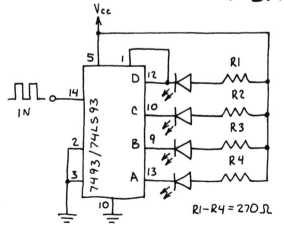

R1-R4 = 270 Ω

COUNTS FROM
0-15 IN BINARY
AND RECYCLES.
GLOWING LED = L
(0); OFF LED = H
(1). 555 TIMER
IC MAKES GOOD
INPUT CLOCK.

TRUTH TABLE

D C B A		D C B A
L L L L	→	H L L L
L L L H		H L L H
L L H L		H L H L
L L H H		H L H H
L H L L		H H L L
L H L H		H H L H
L H H L		H H H L
L H H H		H H H H

BCD UP-DOWN COUNTER 74192

FULLY PROGRAMMABLE BCD COUNTER. OPERATION IS IDENTICAL TO 74193/74LS193 EXCEPT COUNT IS 10-STEP BCD (LLLL–HLLH) INSTEAD OF 16-STEP BINARY. MANY APPLICATIONS FOR 74192/74LS192 AND 74193/74LS193 ARE INTERCHANGEABLE.

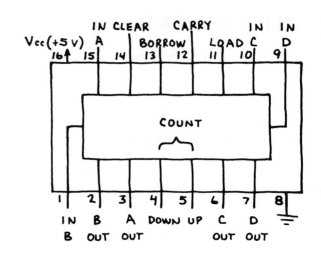

CASCADED COUNTERS

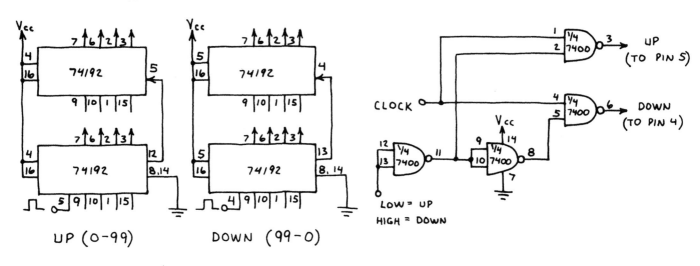

UP (0-99)

DOWN (99-0)

SINGLE UP-DOWN INPUT

LOW = UP
HIGH = DOWN

PROGRAMMABLE COUNT DOWN TIMER

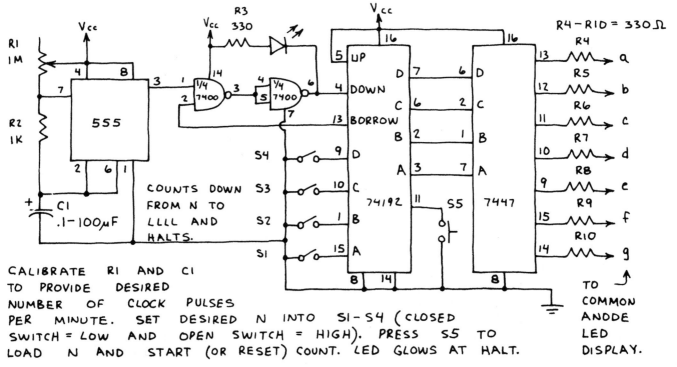

R4-R10 = 330 Ω

COUNTS DOWN FROM N TO LLLL AND HALTS.

TO COMMON ANODE LED DISPLAY.

CALIBRATE R1 AND C1 TO PROVIDE DESIRED NUMBER OF CLOCK PULSES PER MINUTE. SET DESIRED N INTO S1-S4 (CLOSED SWITCH = LOW AND OPEN SWITCH = HIGH). PRESS S5 TO LOAD N AND START (OR RESET) COUNT. LED GLOWS AT HALT.

4-BIT UP COUNTER
74LS161

GENERAL PURPOSE BINARY COUNTER
WITH PROGRAMMABLE INPUTS.
COUNTER ACCEPTS DATA AT INPUTS
WHEN LOAD INPUT GOES LOW.
A LOW AT THE CLEAR INPUT
RESETS THE COUNTER TO LLLL
UPON THE NEXT CLOCK PULSE.
P AND T ARE COUNT ENABLE
INPUTS. BOTH P AND T MUST BE
HIGH TO COUNT. THESE ENABLE
INPUTS ARE NOT AVAILABLE WITH
THE OTHERWISE MORE ADVANCED 74LS193.

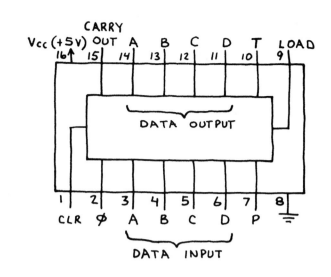

8-BIT COUNTER

RAMP SYNTHESIZER

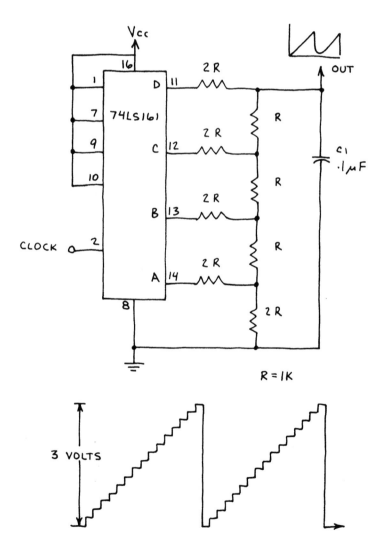

R = 1K

OUTPUT A IS LOWEST ORDER
BIT.

REMOVE C1 TO OBTAIN THIS STAIRCASE.
FREQUENCY OF RAMP AND STAIRCASE
IS 1/16 CLOCK FREQUENCY.

4-BIT UP-DOWN COUNTER 74193/74LS193

VERY VERSATILE 4-BIT COUNTER WITH UP-DOWN CAPABILITY. ANY 4-BIT NUMBER AT THE DCBA INPUTS IS LOADED INTO THE COUNTER WHEN THE LOAD INPUT (PIN 11) IS MADE LOW. THE COUNTER IS CLEARED TO LLLL WHEN THE CLEAR INPUT (PIN 14) IS MADE HIGH. THE BORROW AND CARRY OUTPUTS INDICATE UNDERFLOW OR OVERFLOW BY GOING LOW.

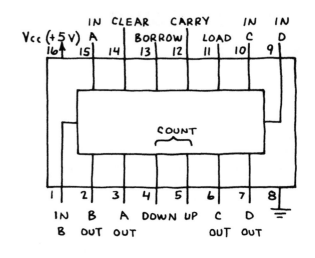

COUNT DOWN FROM N AND RECYCLE

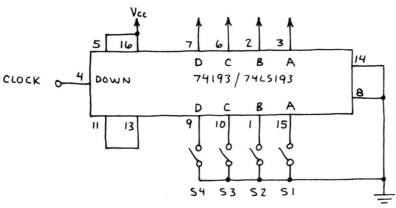

SET DESIRED N INTO S1-S4 (CLOSED SWITCH = LOW AND OPEN SWITCH = HIGH). WHEN COUNT REACHES LLLL AND THEN UNDERFLOWS, THE BORROW PULSE LOADS N AND THE COUNT RECYCLES.

COUNT UP TO N AND HALT

PRESS S1 (NORMALLY CLOSED) TO RESET.

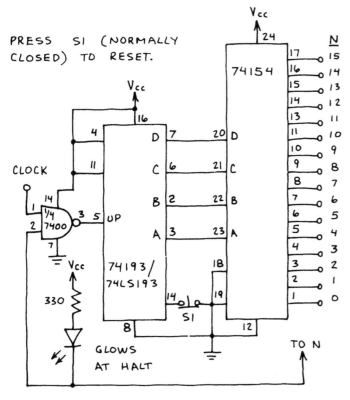

COUNT UP TO N AND RECYCLE

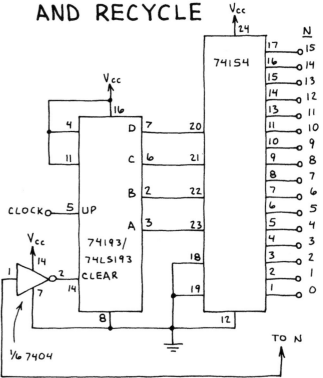

NOTES

4-BIT SHIFT REGISTER 74LS194

BIDIRECTIONAL UNIVERSAL SHIFT REGISTER. SHIFTS RIGHT WHEN SO IS HIGH AND SI IS LOW. SHIFTS LEFT WHEN SO IS LOW AND SI IS HIGH. SHIFTS ONE POSITION PER CLOCK PULSE. LOADS DATA AT INPUTS WHEN SO AND SI ARE HIGH. IMPORTANT: BYPASS POWER SUPPLY PINS WITH $0.1 \mu F$ CAPACITOR!

Vcc (+5v) — DATA OUT —

| 16 | 15 | 14 | 13 | 12 | 11 | 10 | 9 |
| Vcc | A | B | C | D | Ø | SI | SO |

SERIAL INPUTS:
SHIFT LEFT
SHIFT RIGHT

| 1 | 2 | 3 | 4 | 5 | 6 | 7 | 8 |
| CLR | RIGHT IN | A | B | C | D | LEFT IN | |

— DATA IN —

SEQUENCE GENERATOR

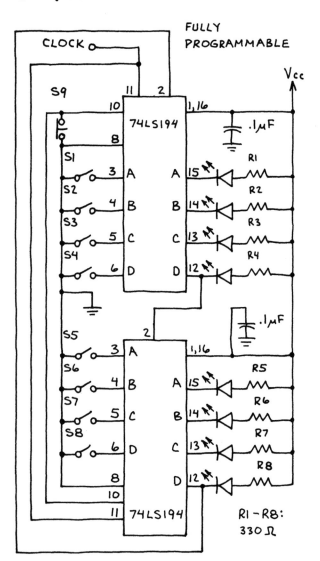

FULLY PROGRAMMABLE

LOAD ANY DESIRED BIT PATTERN INTO S1-S8 (OPEN = HIGH AND CLOSED = LOW). PRESS S9 (NORMALLY CLOSED) TO LOAD. DATA WILL MOVE RIGHT ONE OUTPUT PER CLOCK PULSE. LEDS ARE OPTIONAL.

R1-R8: 330 Ω

BARGRAPH GENERATOR

WHEN POWER IS FIRST APPLIED, MAKE ENABLE INPUT LOW TO START CIRCUIT.

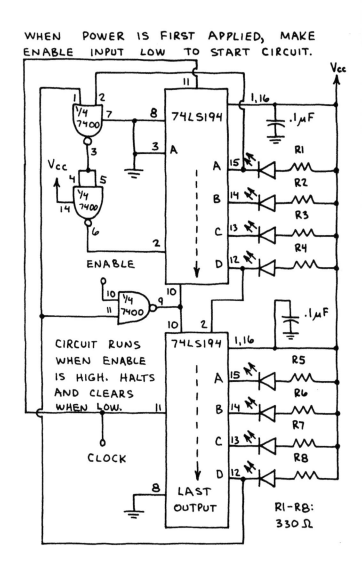

CIRCUIT RUNS WHEN ENABLE IS HIGH. HALTS AND CLEARS WHEN LOW.

R1-R8: 330 Ω

OUTPUTS GO LOW AND STAY LOW ONE AT A TIME FROM LEFT TO RIGHT (A→D) IN SEQUENCE WITH CLOCK. WHEN FINAL OUTPUT GOES LOW, ALL OUTPUTS BUT THE FIRST GO HIGH AND RECYCLE.

8-BIT SHIFT REGISTER
74LS164

DATA AT ONE OF THE TWO SERIAL INPUTS IS ADVANCED ONE BIT FOR EACH CLOCK PULSE. DATA CAN BE EXTRACTED FROM THE 8 PARALLEL OUTPUTS OR IN SERIAL FORM AT ANY SINGLE OUTPUT. ENTER DATA AT EITHER INPUT. THE UNUSED INPUT MUST BE HELD HIGH OR CLOCKING WILL BE INHIBITED. MAKING PIN 9 LOW CLEARS THE REGISTER TO LLLL.

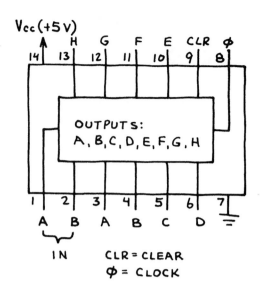

8-BIT SERIAL-TO-PARALLEL DATA CONVERTER

USE FOR RECEIVING BINARY DATA SENT OVER ONE CHANNEL.

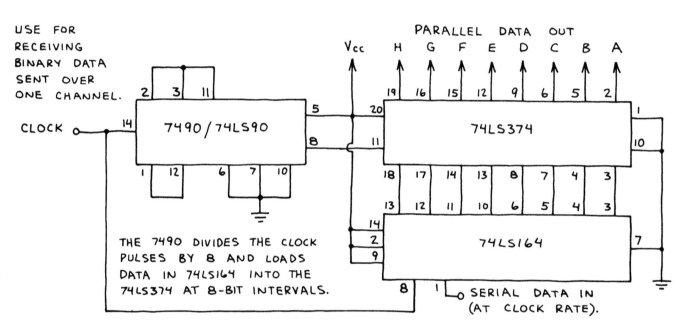

THE 7490 DIVIDES THE CLOCK PULSES BY 8 AND LOADS DATA IN 74LS164 INTO THE 74LS374 AT 8-BIT INTERVALS.

SERIAL DATA IN (AT CLOCK RATE).

PSEUDO-RANDOM VOLTAGE GENERATOR

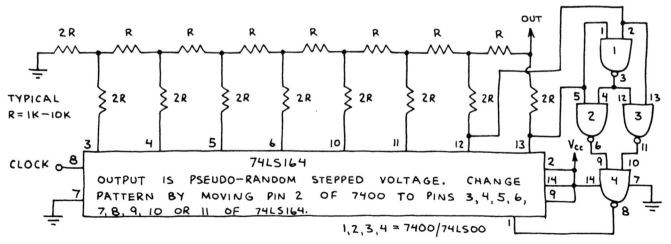

TYPICAL R = 1K – 10K

OUTPUT IS PSEUDO-RANDOM STEPPED VOLTAGE. CHANGE PATTERN BY MOVING PIN 2 OF 7400 TO PINS 3,4,5,6, 7,8,9,10 OR 11 OF 74LS164.

1,2,3,4 = 7400/74LS00

79

OCTAL BUFFER
74LS240

IDEAL FOR INTERFACING
EXTERNAL CIRCUITS TO
HOME COMPUTERS.
INVERTS DATA.

CONTROL (E1, E2)	OUT
L	$\overline{IN}$
H	HI-Z

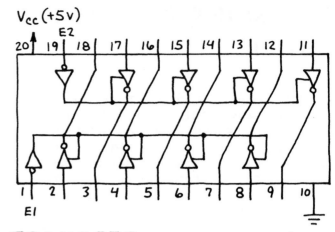

4-BIT BUS TRANSFER

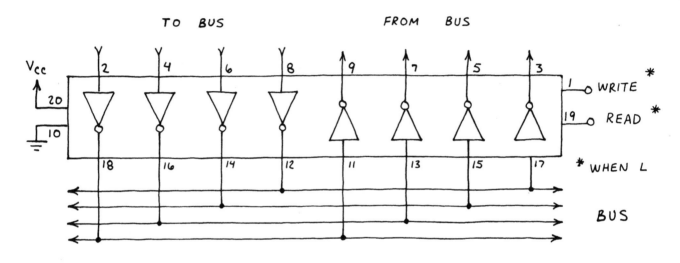

TO BUS

FROM BUS

WRITE *

READ *

* WHEN L

BUS

8-BIT BUS BUFFER

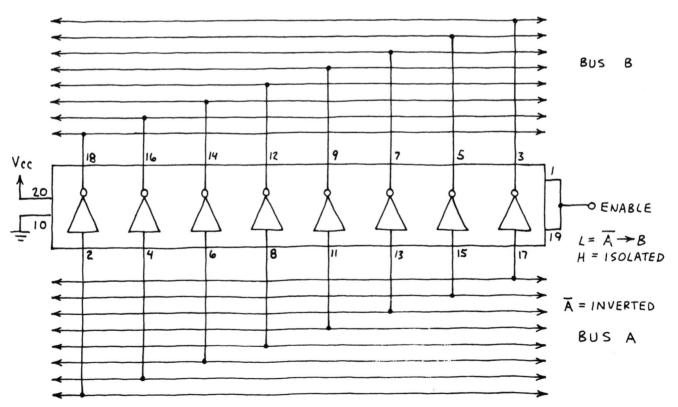

BUS B

ENABLE

$L = \overline{A} \rightarrow B$
H = ISOLATED

$\overline{A}$ = INVERTED

BUS A

OCTAL BUFFER
74LS244

NON-INVERTING VERSION
OF 74LS240. IDEAL FOR
COMPUTER INTERFACING.

CONTROL (E1, E2)	OUT
L	$\overline{IN}$
H	HI-Z

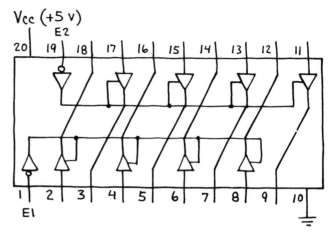

4-BIT BUS TRANSFER

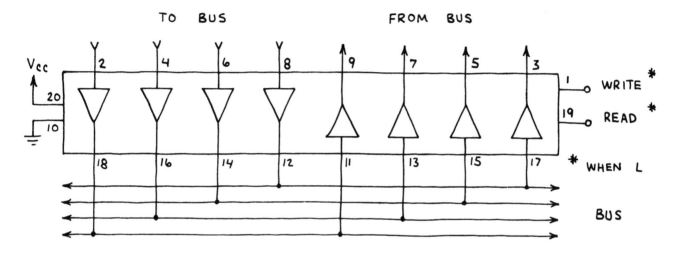

TO BUS   FROM BUS

WRITE *

READ *

* WHEN L

BUS

8-BIT BUS BUFFER

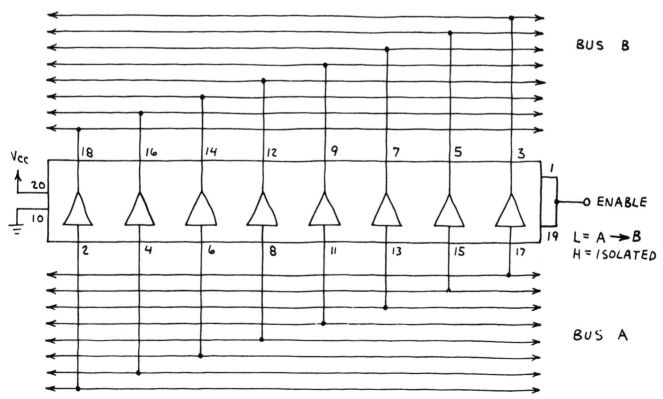

BUS B

ENABLE

L = A → B
H = ISOLATED

BUS A

81

OCTAL D-TYPE LATCH 74LS373

EIGHT "TRANSPARENT" D-TYPE LATCHES. OUTPUT FOLLOWS INPUT WHEN ENABLE IS HIGH. THE DATA AT THE INPUTS IS LOADED WHEN THE ENABLE INPUT IS LOW. THIS CHIP HAS 3-STATE OUTPUTS WHICH ARE CONTROLLED BY PIN I. SEE TRUTH TABLE BELOW.

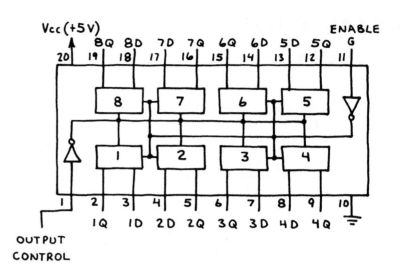

3-STATE REGISTER

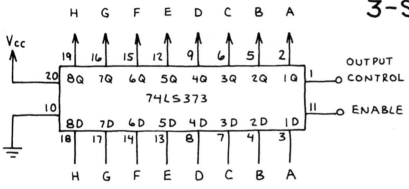

THIS IS A GENERAL PURPOSE 8-BIT STORAGE REGISTER. HERE'S THE TRUTH TABLE:

OUTPUT CONTROL	ENABLE	D	Q
L	H	H	H
L	H	L	L
L	L	X	Q
H	X	X	HI-Z

DATA BUS REGISTERS

H: PLACES OUTPUTS IN HI-Z MODE
L: MAKES DATA AVAILABLE

H: OUTPUTS FOLLOW DATA ON BUS
L: LOAD DATA FROM BUS

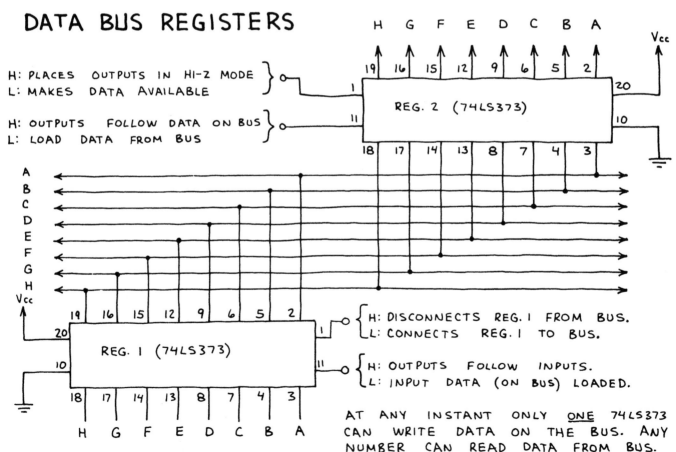

H: DISCONNECTS REG. I FROM BUS.
L: CONNECTS REG. I TO BUS.

H: OUTPUTS FOLLOW INPUTS.
L: INPUT DATA (ON BUS) LOADED.

AT ANY INSTANT ONLY <u>ONE</u> 74LS373 CAN WRITE DATA ON THE BUS. ANY NUMBER CAN READ DATA FROM BUS.

OCTAL D FLIP-FLOP
74LS374

EIGHT D-TYPE EDGE TRIGGERED
FLIP-FLOPS. UNLIKE 74LS373,
OUTPUTS DO <u>NOT</u> FOLLOW
INPUTS. INSTEAD, A RISING
CLOCK PULSE AT PIN 11 LOADS
DATA APPEARING AT INPUTS.
THIS CHIP HAS 3-STATE
OUTPUTS WHICH ARE CONTROLLED
BY PIN 1.

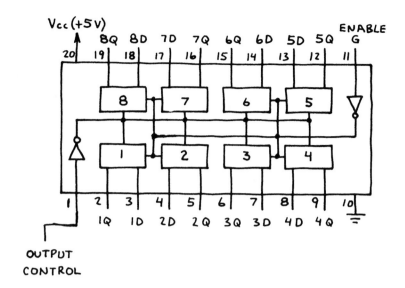

CLOCKED
3-STATE REGISTER

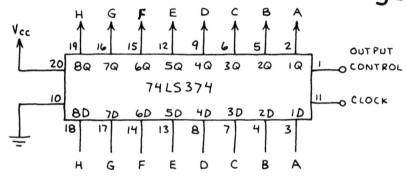

GENERAL PURPOSE
CLOCKED REGISTER.
HERE'S THE TRUTH TABLE:

OUTPUT CONTROL	CLOCK	D	Q
L	⌐	H	H
L	⌐	L	L
L	H	X	Q
H	X	X	HI-Z

COMMON INPUT/OUTPUT BUS REGISTER

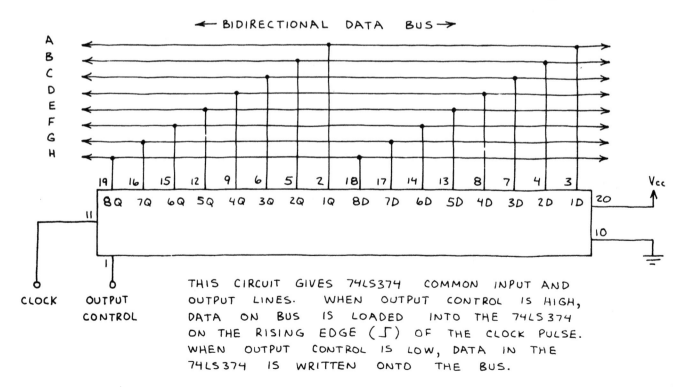

THIS CIRCUIT GIVES 74LS374 COMMON INPUT AND
OUTPUT LINES. WHEN OUTPUT CONTROL IS HIGH,
DATA ON BUS IS LOADED INTO THE 74LS374
ON THE RISING EDGE (⌐) OF THE CLOCK PULSE.
WHEN OUTPUT CONTROL IS LOW, DATA IN THE
74LS374 IS WRITTEN ONTO THE BUS.

83

OCTAL BUS TRANSCEIVER 74LS245

ALLOWS DATA TO BE TRANSFERRED IN EITHER DIRECTION BETWEEN TWO BUSES. INCLUDES HIGH IMPEDANCE (HI-Z) OUTPUTS.

BUS TRANSCEIVER

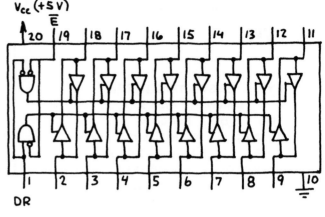

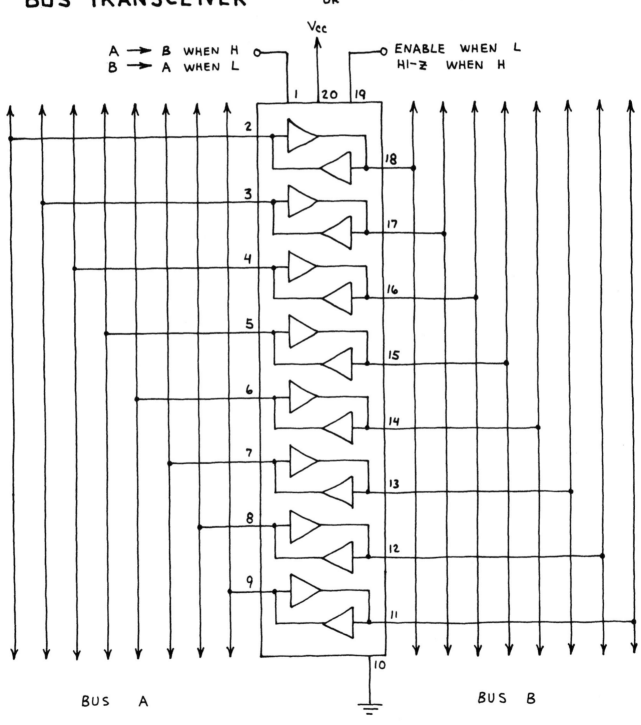

V_{cc} (+5V)

A → B WHEN H
B → A WHEN L

ENABLE WHEN L
HI-Z WHEN H

V_{cc}

BUS A

BUS B

DR

LINEAR INTEGRATED CIRCUITS

INTRODUCTION

THE OUTPUT OF A LINEAR IC IS PROPORTIONAL TO THE SIGNAL AT ITS INPUT. THE CLASSIC LINEAR IC IS THE OPERATIONAL AMPLIFIER. THIS GRAPH SHOWS THE LINEAR INPUT-OUTPUT RELATIONSHIP OF A TYPICAL OP-AMP CIRCUIT:

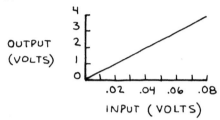

MANY NON-DIGITAL ICs — INCLUDING OP-AMPS — CAN BE USED IN BOTH LINEAR AND NON-LINEAR MODES. THEY ARE SOMETIMES DESCRIBED AS ANALOG ICs.

LINEAR ICs GENERALLY REQUIRE MORE EXTERNAL COMPONENTS THAN DIGITAL ICs. THIS INCREASES THEIR SUSCEPTABILITY TO EXTERNAL NOISE AND MAKES THEM A LITTLE TRICKIER TO USE. ON THE OTHER HAND, SOME LINEAR ICs CAN DO ESSENTIALLY THE SAME THING AS A NETWORK OF DIGITAL CHIPS.

HERE'S A BRIEF DESCRIPTION OF THE LINEAR CHIPS IN THIS SECTION:

VOLTAGE REGULATORS

PROVIDE A STEADY VOLTAGE, EITHER FIXED OR ADJUSTABLE, THAT IS UN-AFFECTED BY CHANGES IN THE SUPPLY VOLTAGE AS LONG AS THE SUPPLY VOLTAGE IS ABOVE THE DESIRED OUTPUT VOLTAGE.

OPERATIONAL AMPLIFIERS

THE IDEAL AMPLIFIER ... ALMOST. HIGH INPUT IMPEDANCE AND GAIN. LOW OUTPUT IMPEDANCE. GAIN IS EASILY CONTROLLED WITH A SINGLE FEEDBACK RESISTOR. FET INPUT OP-AMPS (BIFETS) HAVE A VERY HIGH FREQUENCY RESPONSE. IT'S USUALLY OK TO SUBSTITUTE OP-AMPS IF BOTH ARE NORMALLY POWERED BY A DUAL POLARITY SUPPLY ($\frac{1}{2}$ LF353 FOR 741C, ETC.)... BUT PERFORMANCE WILL IMPROVE OR DECREASE ACCORDING TO THE NEW OP-AMP'S SPECIFICATIONS.

COMPARATOR

SAME AS AN OP-AMP WITHOUT A FEEDBACK RESISTOR. ULTRA-HIGH GAIN GIVES A SNAP-LIKE RESPONSE TO AN INPUT VOLTAGE AT ONE INPUT THAT EXCEEDS A REFERENCE VOLTAGE AT THE SECOND INPUT.

TIMERS

USE ALONE OR WITH OTHER ICs FOR NUMEROUS TIMING AND PULSE GENER-ATION APPLICATIONS.

LED CHIPS

MOST IMPORTANT ARE A FLASHER CHIP AND A DOT-BARGRAPH ANALOG-TO-DIGITAL DISPLAY. VERY EASY TO USE.

OSCILLATORS

A VOLTAGE CONTROLLED OSCILLATOR AND A COMBINED VOLTAGE-TO-FRE-QUENCY AND FREQUENCY-TO-VOLTAGE CONVERTER. ALSO INCLUDED IS A TONE DECODER THAT CAN BE SET TO INDICATE A SPECIFIC FREQUENCY.

AUDIO AMPLIFIERS

THIS SECTION INCLUDES SEVERAL EASY TO USE POWER AMPLIFIERS THAT ARE IDEAL FOR DO-IT-YOURSELF STEREO, PUBLIC ADDRESS SYSTEMS, INTERCOMS AND OTHER AUDIO APPLICATIONS.

VOLTAGE REGULATORS
7805 (5-VOLTS)
7812 (12-VOLTS)
7815 (15-VOLTS)

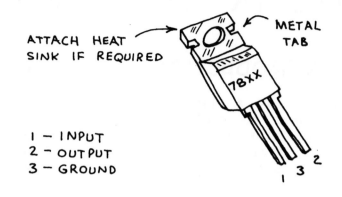

ATTACH HEAT SINK IF REQUIRED

METAL TAB

78XX

1 — INPUT
2 — OUTPUT
3 — GROUND

FIXED VOLTAGE REGULATORS. IDEAL FOR STAND-ALONE POWER SUPPLIES, ON-CARD REGULATORS, AUTOMOBILE BATTERY POWERED PROJECTS, ETC. UP TO 1.5 AMPERES OUTPUT IF PROPERLY HEAT SUNK <u>AND</u> SUFFICIENT INPUT CURRENT AVAILABLE. THERMAL SHUTDOWN CIRCUIT TURNS OFF REGULATOR IF HEATSINK TOO SMALL.

5-VOLT LINE POWERED TTL/LS POWER SUPPLY

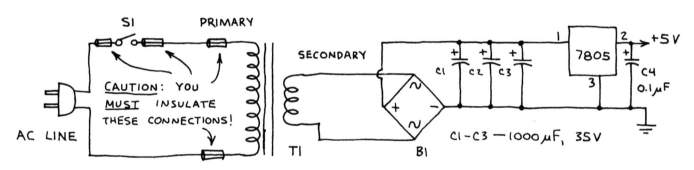

S1   PRIMARY

CAUTION: YOU <u>MUST</u> INSULATE THESE CONNECTIONS!

AC LINE

SECONDARY

C1   C2   C3

7805

+5V

C4
0.1 µF

T1   B1

C1-C3 — 1000 µF, 35V

T1 — 117 — 12.6 V, 1.2 A OR 3A TRANSFORMER
B1 — 1A — 4A FULL WAVE BRIDGE RECTIFIER

VOLTAGE REGULATOR

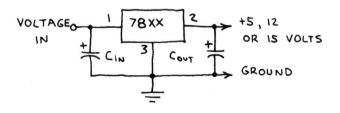

VOLTAGE IN

78XX

C_{IN} C_{OUT}

+5, 12 OR 15 VOLTS

GROUND

C_{IN} — OPTIONAL; USE 0.33 µF OR SO IF REGULATOR FAR FROM POWER SUPPLY.
C_{OUT} — OPTIONAL; USE 0.1 µF OR MORE TO TRAP SPIKES THAT BOTHER LOGIC ICs.

CURRENT REGULATOR

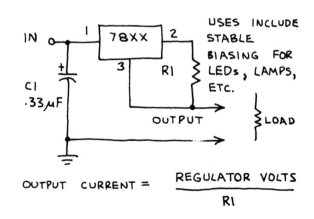

IN

78XX

C1
.33 µF

R1

OUTPUT

LOAD

USES INCLUDE STABLE BIASING FOR LEDs, LAMPS, ETC.

OUTPUT CURRENT = $\dfrac{\text{REGULATOR VOLTS}}{R1}$

-5 VOLT REGULATOR
7905

FIXED -5 VOLT
REGULATOR. CAN BE
USED TO GIVE
ADJUSTABLE VOLTAGE
OUTPUT. UP TO 1.5
AMPERES OUTPUT <u>IF</u>
PROPERLY HEAT SUNK
<u>AND</u> SUFFICIENT INPUT
CURRENT AVAILABLE.
THERMAL SHUTDOWN CIRCUIT
TURNS REGULATOR OFF
IF HEATSINK TOO SMALL.

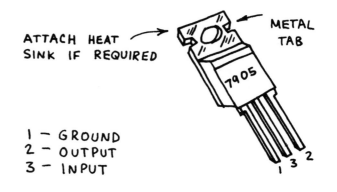

ATTACH HEAT
SINK IF REQUIRED

METAL TAB

1 - GROUND
2 - OUTPUT
3 - INPUT

FIXED -5 VOLT REGULATOR

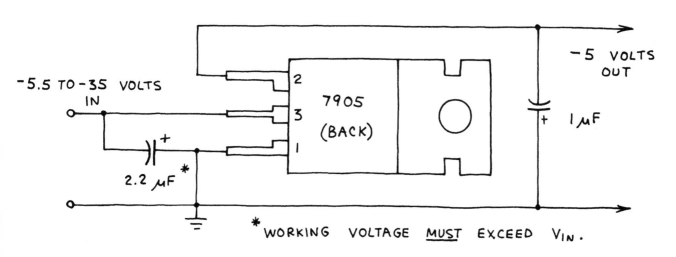

-5.5 TO -35 VOLTS IN

7905 (BACK)

-5 VOLTS OUT

$1 \mu F$

$2.2 \mu F$ *

*WORKING VOLTAGE <u>MUST</u> EXCEED V_{IN}.

ADJUSTABLE NEGATIVE POWER SUPPLY

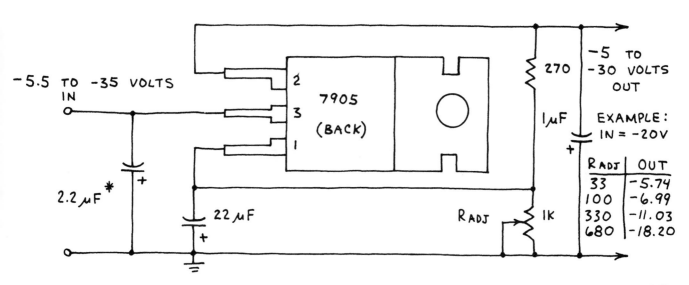

-5.5 TO -35 VOLTS IN

7905 (BACK)

270

-5 TO -30 VOLTS OUT

$1 \mu F$

$2.2 \mu F$ *

$22 \mu F$

R_{ADJ}

1k

EXAMPLE:
IN = -20V

R_{ADJ}	OUT
33	-5.74
100	-6.99
330	-11.03
680	-18.20

87

1.2-37 VOLT REGULATOR
LM317

CAN SUPPLY UP TO 1.5 AMPERES OVER A 1.2-37 VOLT OUTPUT RANGE. NOTE MINIMUM NUMBER OF EXTERNAL COMPONENTS IN BASIC REGULATOR CIRCUIT BELOW. USE HEAT SINK FOR APPLICATIONS REQUIRING FULL POWER OUTPUT. SEE APPROPRIATE DATA BOOK FOR ADDITIONAL INFORMATION:

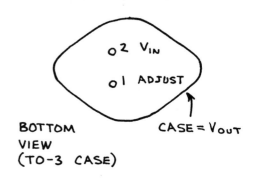

BOTTOM VIEW (TO-3 CASE)

1.25-25 VOLT REGULATOR

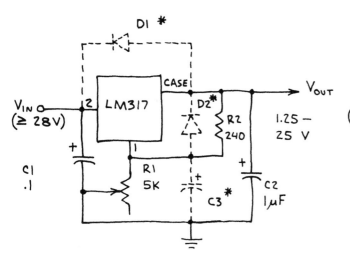

V_{IN} SHOULD BE FILTERED. OK TO OMIT C1 IF V_{IN} VERY CLOSE TO LM317. R1 CONTROLS OUTPUT VOLTAGE.
* ADD IF OUTPUT >25 V AND C2 >25μF.

6-VOLT NICAD CHARGER

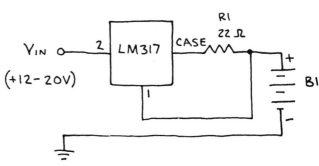

B1 IS BATTERY OF 4 NICKEL CADMIUM STORAGE CELLS IN SERIES. THIS CIRCUIT CHARGES B1 AT A CURRENT OF 51.2 mA. INCREASE R1 TO REDUCE CURRENT. FOR EXAMPLE, CURRENT IS 43 mA WHEN R1 IS 24 OHMS.

PROGRAMMABLE POWER SUPPLY

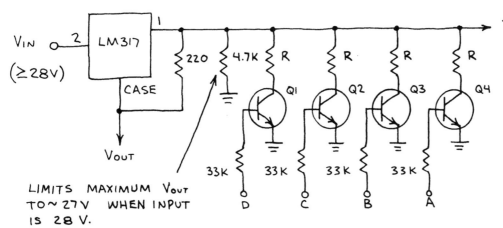

TO ADDITIONAL STAGES

DCBA INPUTS: CONNECT TO PIN 2 TO SELECT.

R	V_{OUT}
100	1.8
330	3.0
470	4.0
1K	7.3
2.2K	13.5
3.3K	18.0

LIMITS MAXIMUM V_{OUT} TO ~27V WHEN INPUT IS 28 V.

88

-1.2 TO -37 VOLT REGULATOR
337T

CAN SUPPLY UP TO -1.5
AMPERES OVER A -1.2
TO -37 VOLT OUTPUT
RANGE. FEW EXTERNAL
COMPONENTS REQUIRED.
COMPLEMENTS LM317
ADJUSTABLE POSITIVE
REGULATOR.

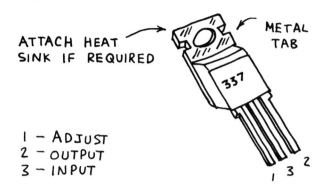

ATTACH HEAT
SINK IF REQUIRED

METAL TAB

1 – ADJUST
2 – OUTPUT
3 – INPUT

ADJUSTABLE NEGATIVE REGULATOR

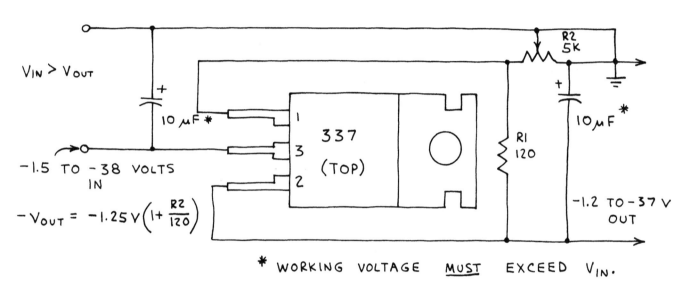

$V_{IN} > V_{OUT}$

10 μF *

-1.5 TO -38 VOLTS
IN

$-V_{OUT} = -1.25V\left(1 + \dfrac{R2}{120}\right)$

R2
5K

R1
120

10 μF *

-1.2 TO -37 V
OUT

* WORKING VOLTAGE <u>MUST</u> EXCEED V_{IN}.

PRECISION LED REGULATOR

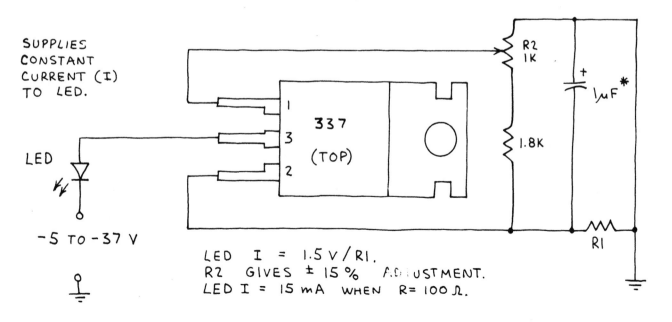

SUPPLIES
CONSTANT
CURRENT (I)
TO LED.

LED

-5 TO -37 V

R2
1K

1 μF *

1.8K

R1

LED I = 1.5 V / R1.
R2 GIVES ± 15 % ADJUSTMENT.
LED I = 15 mA WHEN R = 100 Ω.

2-37 VOLT REGULATOR 723

VERY VERSATILE SERIES
REGULATOR. UP TO 40 VOLTS
INPUT AND 2-37 VOLT OUTPUT.
MAXIMUM OUTPUT CURRENT OF
150 mA CAN BE EXTENDED TO
10 A BY ADDING EXTERNAL
POWER TRANSISTORS. SHOWN
BELOW ARE TWO BASIC
CIRCUITS. TRY THESE, THEN
SEE APPROPRIATE DATA BOOK
FOR ADDITIONAL CIRCUITS.

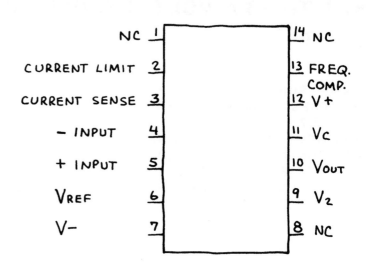

2-7 VOLT REGULATOR

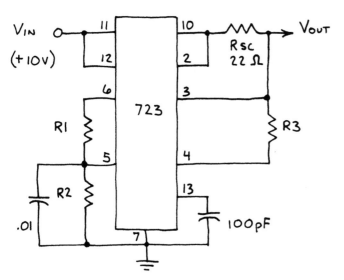

TYPICAL VALUES

V_{OUT}	R1	R2	R3
3.0	4.12 K	3.01 k	1.74K
3.6	3.57 K	3.65 K	1.80K
5.0	2.15 K	4.99 K	1.50K
6.0	1.15 K	6.04 K	966

FOR ANY VOLTAGE BETWEEN 2-7
VOLTS:

$$V_{out} = \left(V_{REF}^{*}\right) \times \left(\frac{R2}{R + R2}\right)$$

$^{*}V_{REF} = 6.8 - 7.5$ V (MEASURE AT PIN 6)

7-37 VOLT REGULATOR

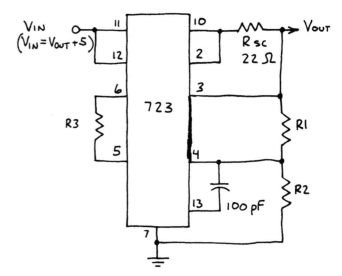

TYPICAL VALUES

V_{OUT}	R1	R2	R3
9	1.87 k	7.15 k	.48K
12	4.87 k	7.15 k	2.90k
15	7.87 k	7.15 k	3.75 k
28	21.0 k	7.15 k	5.33 k

FOR ANY VOLTAGE BETWEEN 7-37
VOLTS:

$$V_{out} = \left(V_{REF}^{*}\right) \times \left(\frac{R1 + R2}{R2}\right)$$

$R3 = \frac{R1 \times R2}{R1 + R2}$ (R3, WHICH IS OPTIONAL, GIVES TEMPERATURE STABILITY)

90

ADJUSTABLE SHUNT (ZENER) REGULATOR
TL431

EASY TO USE THREE
TERMINAL ADJUSTABLE
PRECISION SHUNT
REGULATOR. OUTPUT
CAN BE SET TO FROM
2.5 TO 36 VOLTS.

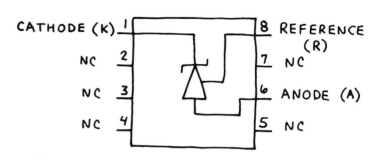

ADJUSTABLE REGULATOR

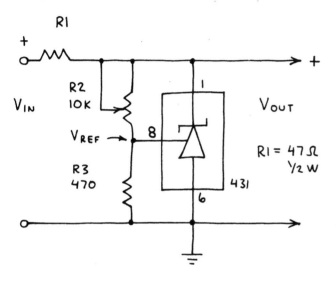

$$V_{OUT} = (1 + R1/R2)\ V_{REF} = 3-30V$$

VOLTAGE DETECTOR

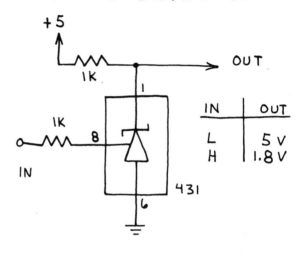

IN	OUT
L	5 V
H	1.8 V

USE TO DETECT
TTL LOGIC LEVELS.

SIMPLE TIMER

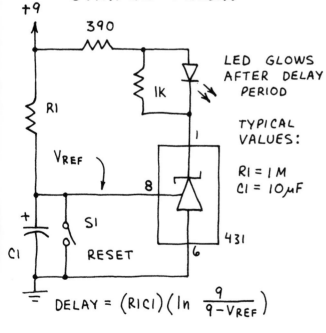

LED GLOWS
AFTER DELAY
PERIOD

TYPICAL
VALUES:

R1 = 1M
C1 = 10μF

$$DELAY = (R1C1)\left(\ln \frac{9}{9-V_{REF}}\right)$$

1.5 TO 5 V POWER SUPPLY

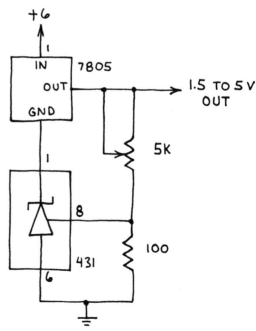

91

1.2 TO 33 VOLT REGULATOR
350T

CAN SUPPLY UP TO
3 AMPERES OVER 1.2
TO 33 VOLT OUTPUT
RANGE. FEW EXTERNAL
COMPONENTS REQUIRED.
HEAT SINK REQUIRED
FOR FULL POWER OUTPUT.

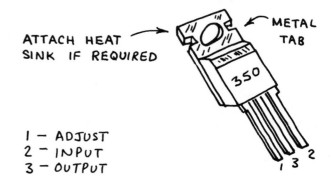

ATTACH HEAT
SINK IF REQUIRED

METAL
TAB

1 — ADJUST
2 — INPUT
3 — OUTPUT

1.2 TO 20 VOLT REGULATOR

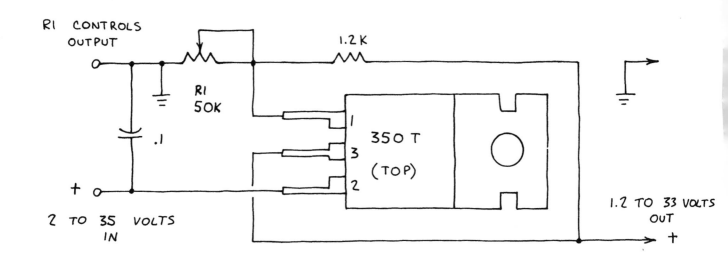

R1 CONTROLS
OUTPUT

1.2 K

R1
50K

.1

350 T
(TOP)

2 TO 35 VOLTS
IN

1.2 TO 33 VOLTS
OUT
+

POWER PULSE GENERATOR

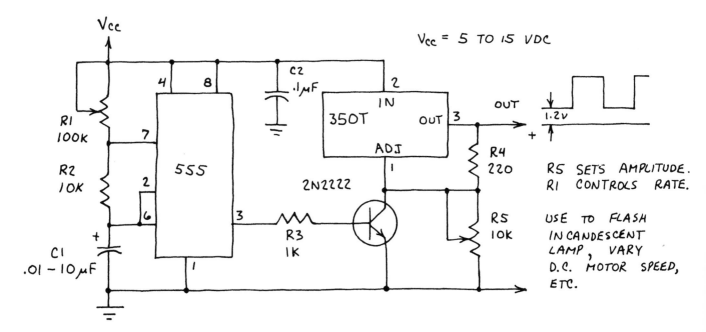

Vcc

Vcc = 5 TO 15 VDC

R1
100K

R2
10K

C1
.01 – 10 μF

555

C2
.1 μF

350T

IN

OUT

ADJ

2N2222

R3
1K

OUT

1.2 V

R4
220

R5
10K

R5 SETS AMPLITUDE.
R1 CONTROLS RATE.

USE TO FLASH
INCANDESCENT
LAMP, VARY
D.C. MOTOR SPEED,
ETC.

OPERATIONAL AMPLIFIER 741C

THE MOST POPULAR OP-AMP. USE FOR ALL GENERAL PURPOSE APPLICATIONS. (FOR SINGLE SUPPLY OPERATION AND VERY HIGH INPUT IMPEDANCE, USE OTHER OP-AMPS IN THIS NOTEBOOK.)

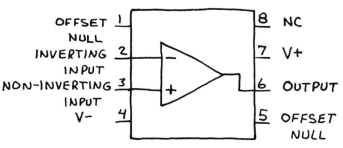

INVERTING AMPLIFIER

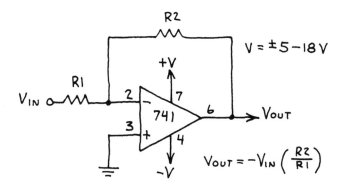

$V = \pm 5 - 18V$

$$V_{OUT} = -V_{IN}\left(\frac{R2}{R1}\right)$$

NON-INVERTING AMPLIFIER

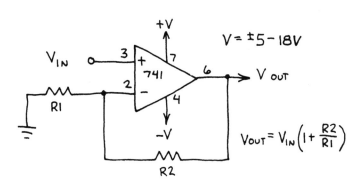

$V = \pm 5 - 18V$

$$V_{OUT} = V_{IN}\left(1 + \frac{R2}{R1}\right)$$

UNITY GAIN FOLLOWER

USE TO COUPLE HIGH IMPEDANCE TO LOW IMPEDANCE.

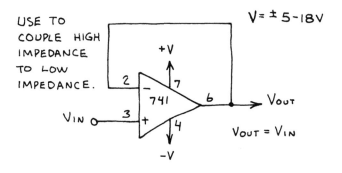

$V = \pm 5 - 18V$

$V_{OUT} = V_{IN}$

COMPARATOR

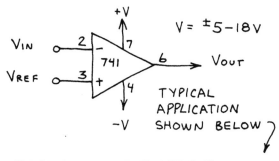

$V = \pm 5 - 18V$

TYPICAL APPLICATION SHOWN BELOW

LEVEL DETECTOR

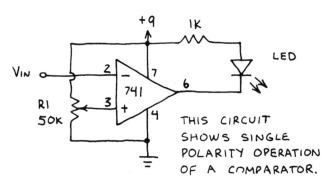

THIS CIRCUIT SHOWS SINGLE POLARITY OPERATION OF A COMPARATOR.

SINGLE POLARITY SUPPLY

$+V = 5 - 18V$

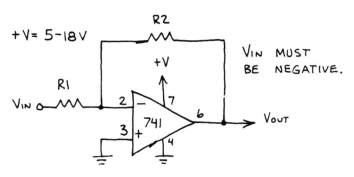

V_{IN} MUST BE NEGATIVE.

TYPICAL USES: AMPLIFICATION OF DC VOLTAGE AND PULSES.

R1 SETS THE VOLTAGE DETECTION THRESHOLD (UP TO +9). WHEN V_{IN} EXCEEDS THE THRESHOLD (ALSO CALLED THE REFERENCE), THE LED GLOWS.

93

OPERATIONAL AMPLIFIER (CONTINUED)
741C

BASIC INTEGRATOR

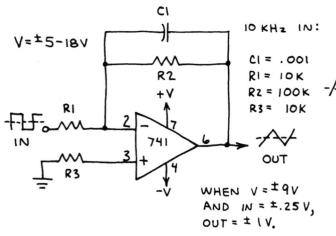

$V = \pm 5 - 18V$

10 KHz IN:

$C1 = .001$
$R1 = 10K$
$R2 = 100K$
$R3 = 10K$

WHEN $V = \pm 9V$
AND IN $= \pm .25V$,
OUT $= \pm 1V$.

BASIC DIFFERENTIATOR

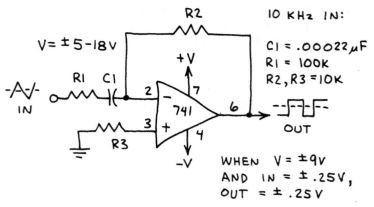

$V = \pm 5 - 18V$

10 KHz IN:

$C1 = .00022 \mu F$
$R1 = 100K$
$R2, R3 = 10K$

WHEN $V = \pm 9V$
AND IN $= \pm .25V$,
OUT $= \pm .25V$

CLIPPING AMPLIFIER

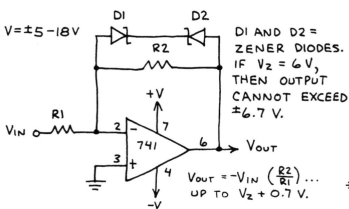

$V = \pm 5 - 18V$

D1 AND D2 =
ZENER DIODES.
IF $V_z = 6V$,
THEN OUTPUT
CANNOT EXCEED
$\pm 6.7V$.

$V_{OUT} = -V_{IN} \left(\frac{R2}{R1} \right) \ldots$
UP TO $V_z + 0.7V$.

BRIDGE AMPLIFIER

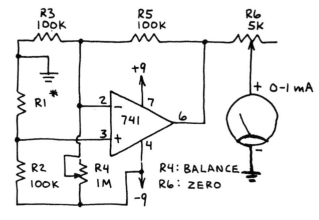

R4: BALANCE
R6: ZERO

* R1 IS UNKNOWN RESISTOR. USE CdS
CELL FOR R1 TO MAKE A <u>VERY</u>
SENSITIVE LIGHT METER.

SUMMING AMPLIFIER

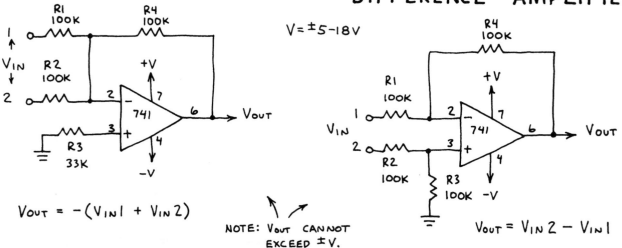

$V_{OUT} = -(V_{IN}1 + V_{IN}2)$

NOTE: V_{OUT} CANNOT
EXCEED $\pm V$.

DIFFERENCE AMPLIFIER

$V = \pm 5 - 18V$

$V_{OUT} = V_{IN}2 - V_{IN}1$

OPERATIONAL AMPLIFIER (CONTINUED)
741C

LIGHT WAVE RECEIVER

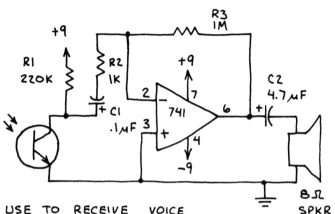

USE TO RECEIVE VOICE MODULATED LIGHT WAVES. OK TO USE SINGLE POLARITY POWER SUPPLY FOR NON-VOICE RECEPTION.

60-Hz NOTCH FILTER

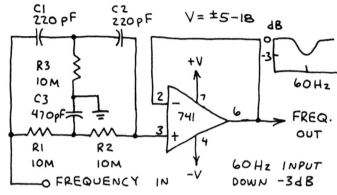

60 Hz INPUT DOWN -3 dB

HIGH PASS ACTIVE FILTER

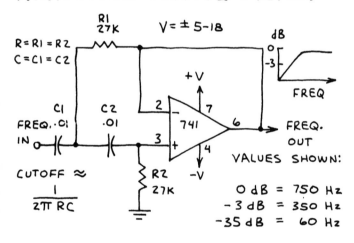

$R = R1 = R2$
$C = C1 = C2$

$$CUTOFF \approx \frac{1}{2\pi RC}$$

VALUES SHOWN:

0 dB	=	750 Hz
-3 dB	=	350 Hz
-35 dB	=	60 Hz

LOW PASS ACTIVE FILTER

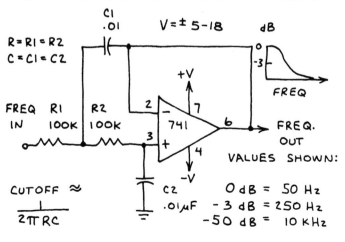

$R = R1 = R2$
$C = C1 = C2$

$$CUTOFF \approx \frac{1}{2\pi RC}$$

VALUES SHOWN:

0 dB	=	50 Hz
-3 dB	=	250 Hz
-50 dB	=	10 kHz

4-BIT D/A CONVERTER

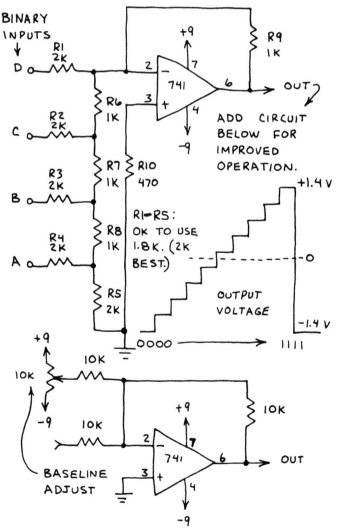

BINARY INPUTS

ADD CIRCUIT BELOW FOR IMPROVED OPERATION.

R1-R5: OK TO USE 1.8K. (2K BEST.)

OUTPUT VOLTAGE

BASELINE ADJUST

OPERATIONAL AMPLIFIER (CONTINUED)
741C

OPTICAL POWER METER

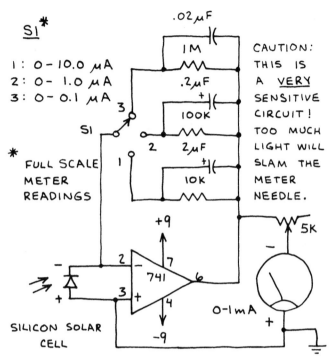

S1*

1: 0 - 10.0 μA
2: 0 - 1.0 μA
3: 0 - 0.1 μA

* FULL SCALE METER READINGS

CAUTION: THIS IS A VERY SENSITIVE CIRCUIT! TOO MUCH LIGHT WILL SLAM THE METER NEEDLE.

SILICON SOLAR CELL

THIS CIRCUIT CAN BE USED AS A FAIRLY GOOD QUALITY RADIOMETER.

BARGRAPH LIGHT METER

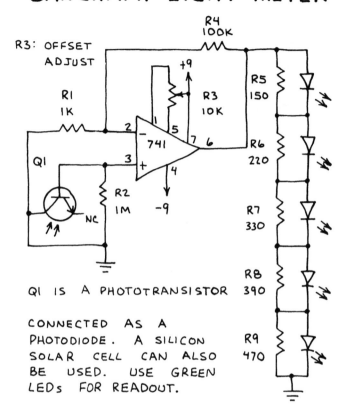

QI IS A PHOTOTRANSISTOR

CONNECTED AS A PHOTODIODE. A SILICON SOLAR CELL CAN ALSO BE USED. USE GREEN LEDs FOR READOUT.

ELECTRONIC BELL

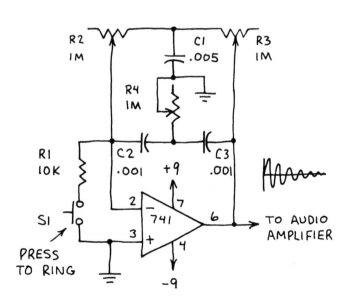

ADJUST R3 TO JUST BELOW OSCILLATION POINT. ADJUST R2 AND R3 FOR SOUNDS SUCH AS BELL, DRUM, TINKLING, ETC.

AUDIBLE LIGHT SENSOR

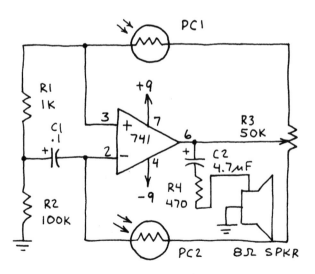

PC1, PC2 — CdS PHOTOCELLS

LIGHT ON PC1 DECREASES TONE FREQUENCY.
LIGHT ON PC2 INCREASES TONE FREQUENCY.

DUAL OPERATIONAL AMPLIFIER 1458

TWO 741C OP-AMPS IN A
SINGLE 8-PIN MINI-DIP. TRY
TO USE THIS CHIP FOR
CIRCUITS THAT REQUIRE TWO
OR MORE 741's. YOU'LL SAVE
TIME, SPACE AND MONEY.

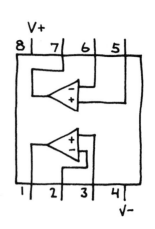

PEAK DETECTOR

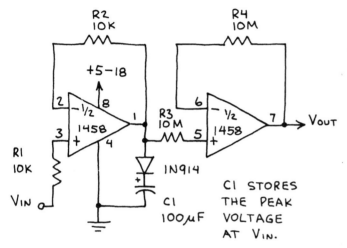

C1 STORES
THE PEAK
VOLTAGE
AT V_{IN}.

APPLICATIONS INCLUDE USE AS
ANALOG "MEMORY" THAT STORES
PEAK AMPLITUDE OF A FLUCTUATING
VOLTAGE.

PULSE GENERATOR

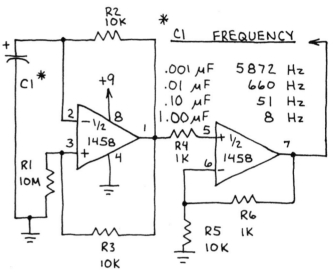

C1	FREQUENCY
.001 µF	5872 Hz
.01 µF	660 Hz
.10 µF	51 Hz
1.00 µF	8 Hz

PULSES ARE DC. AMPLITUDE
WHEN C1 = 0.1 µF IS 5 VOLTS.

FUNCTION GENERATOR

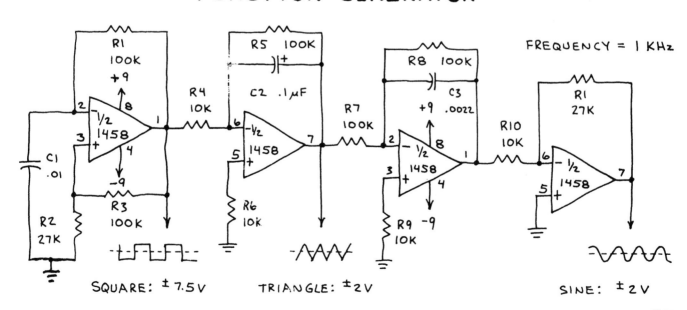

FREQUENCY = 1 KHz

SQUARE: ±7.5V   TRIANGLE: ±2V   SINE: ±2V

97

DUAL OPERATIONAL AMPLIFIER LF353N (JFET INPUT)

HIGH IMPEDANCE (10^{12} OHM) JUNCTION FET INPUTS. OUTPUT SHORT CIRCUIT PROTECTION. HIGH SLEW RATE (13 V/μsec), LOW NOISE OPERATION. AMPLIFIERS ARE SIMILAR TO THOSE IN THE TL084C. NOTE THAT PIN CONNECTIONS ARE THE SAME AS 1458. THIS OP-AMP, HOWEVER, OFFERS MUCH BETTER PERFORMANCE.

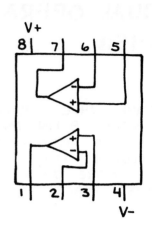

SAMPLE AND HOLD

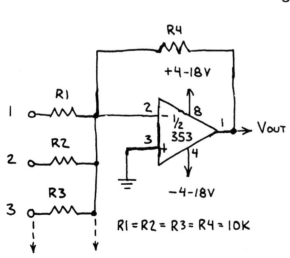

* OPTIONAL

S/H: H = SAMPLE
     L = HOLD

PEAK DETECTOR

TRACKS V_{IN} AND STORES PEAK V_{IN} IN C1.

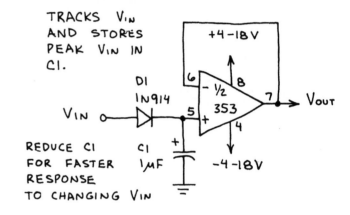

REDUCE C1 FOR FASTER RESPONSE TO CHANGING V_{IN}

PROGRAMMABLE GAIN OP-AMP

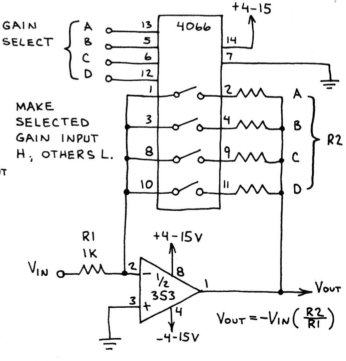

MAKE SELECTED GAIN INPUT H, OTHERS L.

$$V_{OUT} = -V_{IN}\left(\frac{R2}{R1}\right)$$

AUDIO MIXER

$R1 = R2 = R3 = R4 = 10K$

CONNECT OUTPUTS OF PREAMPLIFIERS TO INPUTS 1-3. OK TO ADD MORE CHANNELS. WORKS WELL WITH TL084 MICROPHONE PREAMPLIFIERS.

QUAD OPERATIONAL AMPLIFIER
TLO84C (JFET INPUT)

HIGH IMPEDANCE (10^{12} OHMS) JUNCTION
FET INPUTS. OUTPUT SHORT CIRCUIT
PROTECTION. HIGH SLEW RATE (12 V/
μSEC) PLUS LOW NOISE OPERATION.
PERFORMANCE SIMILAR TO LF353N.
NOTE THAT PIN CONNECTIONS ARE
SAME AS LM324.

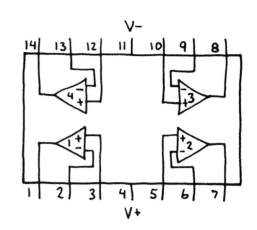

MICROPHONE PREAMPLIFIER

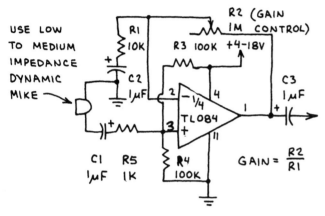

USE LOW
TO MEDIUM
IMPEDANCE
DYNAMIC
MIKE

$$\text{GAIN} = \frac{R2}{R1}$$

NOTE SINGLE POLARITY POWER SUPPLY
(THANKS TO R3 AND R4) AND AC
COUPLING.

LOW-Z PREAMPLIFIER

USE LOW
IMPEDANCE
(LOW-Z)
MICROPHONE

OK TO USE
8Ω SPKR AS
MICROPHONE.
CONNECT DIRECTLY
TO INPUTS (POOR
TO FAIR) OR USE
TRANSFORMER (GOOD):

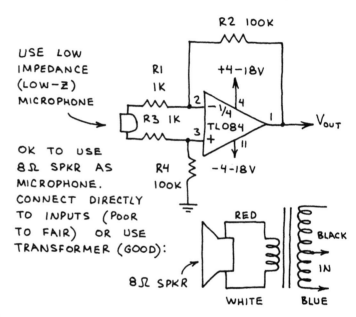

INFRARED VOICE COMMUNICATOR

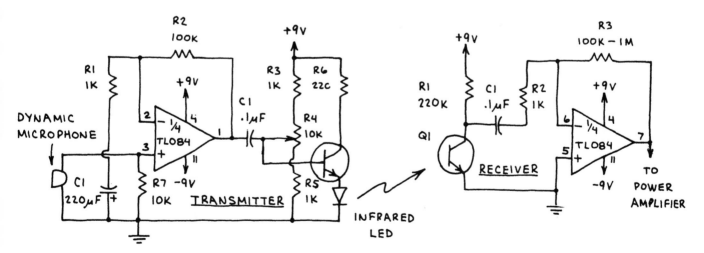

POINT THE LED AT Q1 AND ADJUST R4 UNTIL
BEST VOICE QUALITY IS OBTAINED. (R4 APPLIES
PREBIAS TO LED.) R6 LIMITS MAXIMUM LED
CURRENT TO A SAFE 40 mA.

MAXIMUM RANGE: HUNDREDS
OF FEET AT NIGHT WITH
LENSES AT Q1 AND LED.
POWER AMP: SEE LM386.

QUAD OPERATIONAL AMPLIFIER
LM324N

OPERATES FROM SINGLE POLARITY
POWER SUPPLY. MORE GAIN (100 dB)
BUT LESS BANDWIDTH (1 MHz WHEN
GAIN IS 1) THAN THE LM3900 QUAD
OP-AMP. NOTE UNUSUAL LOCATION
OF POWER SUPPLY PINS. CAUTION:
SHORTING THE OUTPUTS DIRECTLY
TO V+ OR GND OR REVERSING THE
POWER SUPPLY MAY DAMAGE THIS CHIP.

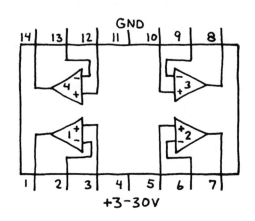

+3-30V

BANDPASS FILTER

BANDPASS
FREQUENCY:
1 KHz

INFRARED TRANSMITTER

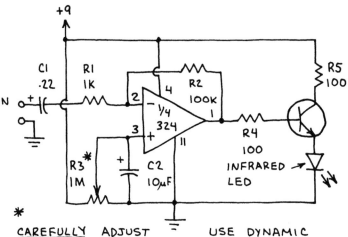

* CAREFULLY ADJUST
R3 FOR BEST VOICE
QUALITY. FOR MORE
POWER REDUCE R5
TO 50Ω... BUT DO
NOT ALLOW MORE THAN
30 mA THROUGH LED!

USE DYNAMIC
MICROPHONE AT
INPUT. RECEIVE
SIGNAL WITH
PHOTOTRANSISTOR
PLUS OP-AMP.

PULSE GENERATOR

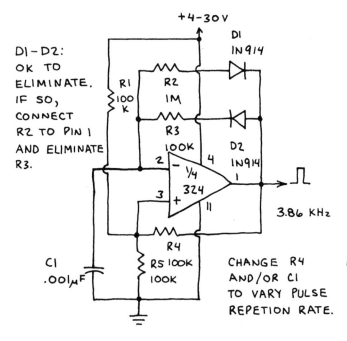

D1-D2:
OK TO
ELIMINATE.
IF SO,
CONNECT
R2 TO PIN 1
AND ELIMINATE
R3.

3.86 KHz

CHANGE R4
AND/OR C1
TO VARY PULSE
REPETION RATE.

INTERFACE CIRCUITS

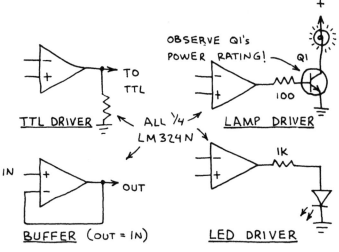

OBSERVE Q1's
POWER RATING!

TTL DRIVER

ALL ¼
LM324N

LAMP DRIVER

BUFFER (OUT = IN)

LED DRIVER

QUAD OPERATIONAL AMPLIFIER
LM3900N

OPERATES FROM SINGLE POLARITY
POWER SUPPLY. LESS GAIN (70 dB)
BUT WIDER BANDWIDTH (25 MHz AT
GAIN OF 1) THAN THE LM324 QUAD
OP-AMP. NOTE STANDARD POWER
SUPPLY PIN LOCATIONS. CAUTION:
SHORTING THE OUTPUTS DIRECTLY TO V+
OR GROUND OR REVERSED POWER
CONNECTIONS MAY DAMAGE THIS CHIP.

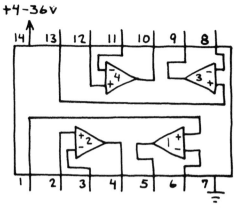

NOTE: DO NOT SUBSTITUTE
LM3900 FOR OTHER OP-AMPS.

ASTABLE MULTIVIBRATOR

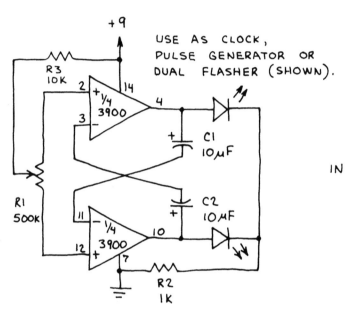

USE AS CLOCK,
PULSE GENERATOR OR
DUAL FLASHER (SHOWN).

TOGGLE FLIP-FLOP

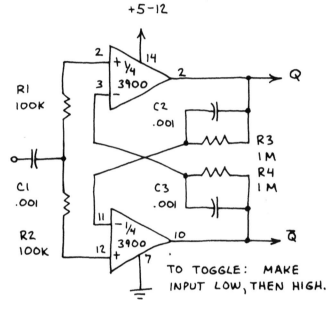

TO TOGGLE: MAKE
INPUT LOW, THEN HIGH.

FUNCTION GENERATOR

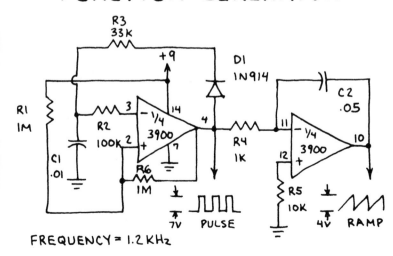

FREQUENCY = 1.2 kHz

X10 AMPLIFIER

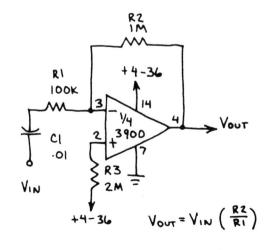

$$V_{OUT} = V_{IN} \left(\frac{R2}{R1} \right)$$

QUAD COMPARATOR
LM339

FOUR INDEPENDENT VOLTAGE COMPARATORS
IN A SINGLE PACKAGE. NOTE THAT
A SINGLE POLARITY POWER SUPPLY
IS REQUIRED. (MOST COMPARATORS ARE
DESIGNED PRIMARILY FOR DUAL SUPPLY
OPERATION.) NOTE UNUSUAL LOCATION OF THE
SUPPLY PINS. COMPARATORS MAY OSCILLATE
IF OUTPUT LEAD IS TOO CLOSE TO INPUT LEADS.
GROUND ALL PINS OF UNUSED COMPARATORS.

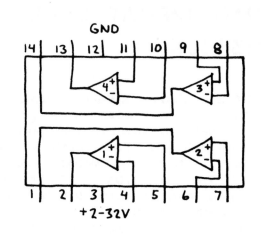

NON-INVERTING COMPARATOR

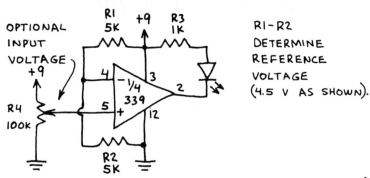

R1-R2
DETERMINE
REFERENCE
VOLTAGE
(4.5 V AS SHOWN).

LED GLOWS WHEN INPUT VOLTAGE (PIN 5)
FALLS BELOW REFERENCE VOLTAGE (PIN 4).

INVERTING COMPARATOR

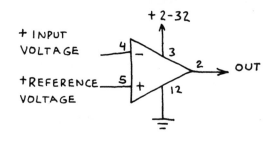

INVERTING COMPARATOR
WITH HYSTERESIS

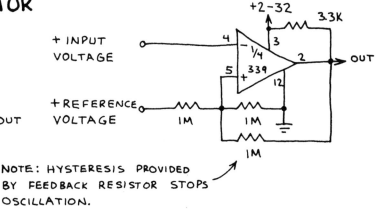

NON-INVERTING COMPARATOR
WITH HYSTERESIS

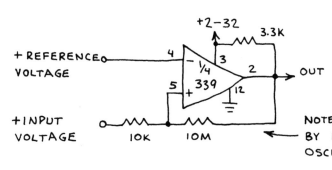

NOTE: HYSTERESIS PROVIDED
BY FEEDBACK RESISTOR STOPS
OSCILLATION.

TTL DRIVER

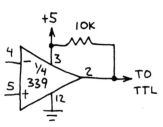

CMOS DRIVER

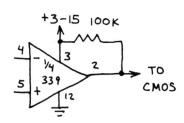

3-STATE OUTPUT

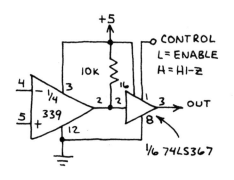

QUAD COMPARATOR (CONTINUED)
LM339

LED BARGRAPH READOUT

WINDOW COMPARATOR

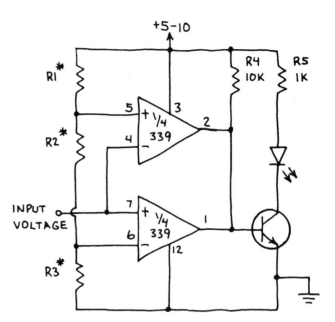

THE LED GLOWS WHEN THE INPUT VOLTAGE IS WITHIN THE WINDOW DETERMINED BY R1-R3. THE WINDOW IS 4-8 MILLIVOLTS WIDE *WHEN R1= 500Ω, R2=1200Ω AND R3= 1M. IT EXTENDS FROM 1.5 -4.2 VOLTS WHEN R1 AND R3= 15,000Ω AND R2= 25,000Ω. USE POTS FOR R1-R3 FOR A FULLY ADJUSTABLE WINDOW.

ADJUST R1 TO ACHIEVE SENSITIVITY UP TO A FEW MILLIVOLTS PER LED. SEE POPULAR ELECTRONICS (SEPT. 1978, pp. 92-97).

PROGRAMMABLE LIGHT METER

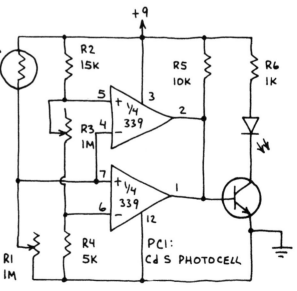

PC1: CdS PHOTOCELL

ADJUST R1 AND R3 SO LED GLOWS WHEN LIGHT AT PC1 IS ABOVE OR BELOW ANY DESIRED LEVEL.

SQUAREWAVE OSCILLATOR

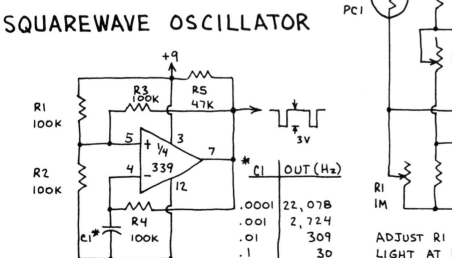

C1	OUT (Hz)
.0001	22,078
.001	2,724
.01	309
.1	30
1.0	4

LED FLASHER/OSCILLATOR
3909

EASIEST TO USE IC IN THIS
NOTEBOOK. FLASHES LEDs OR
CAN BE USED AS TONE SOURCE.
WILL DRIVE SPEAKER DIRECTLY.
WILL FLASH A RED LED WHEN V+
IS ONLY 1.3 VOLTS.

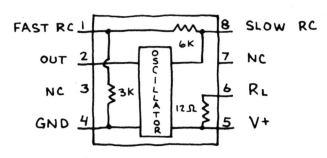

LED FLASHER

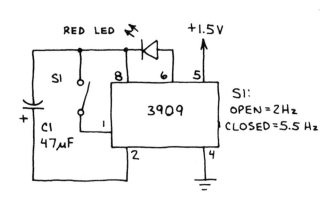

S1:
OPEN = 2Hz
CLOSED = 5.5 Hz

POWER FLASHER

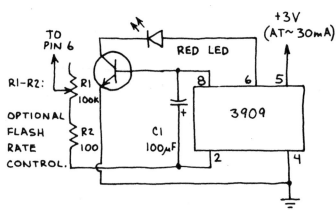

INFRARED TRANSMITTERS

TRANSMITS
STEADY
1 KHz
TONE.

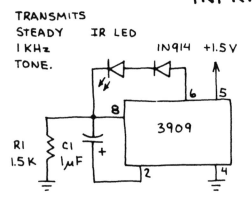

TRANSMITS
DISTINCTIVE
WARBLE TONE

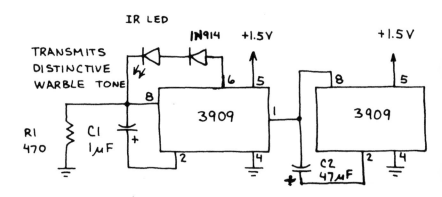

LIGHT CONTROLLED TONE

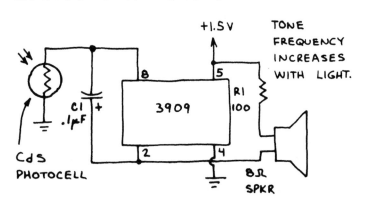

TONE
FREQUENCY
INCREASES
WITH LIGHT.

LAMP FLASHER

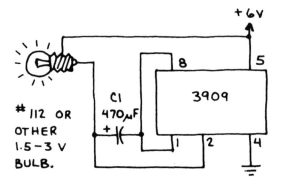

104

LED FLASHER/OSCILLATOR (CONTINUED)
3909

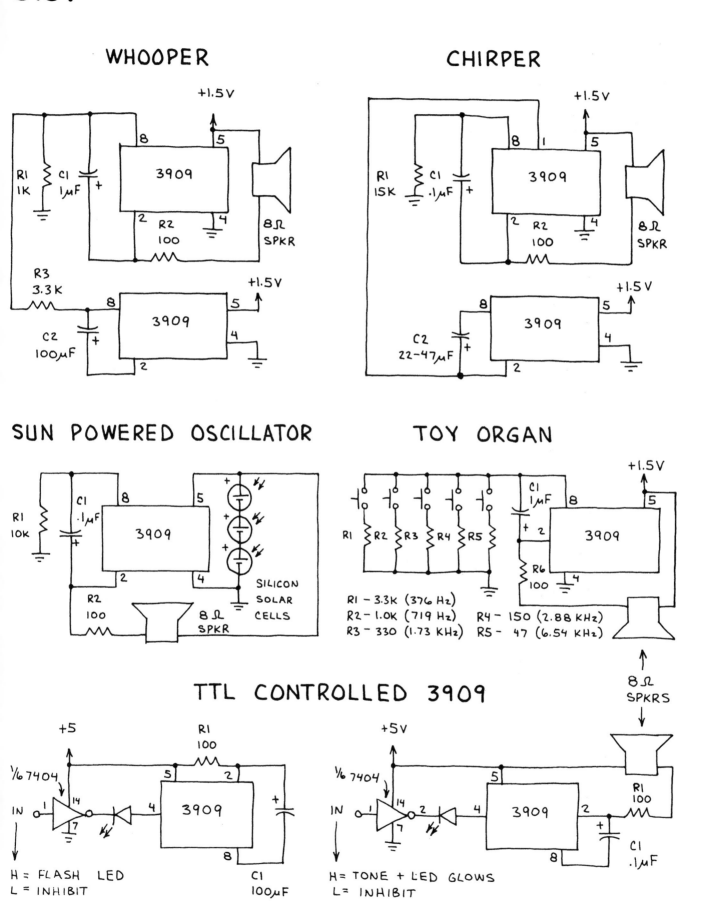

WHOOPER

R1 1K
C1 1μF
3909
R2 100
+1.5V
8Ω SPKR

R3 3.3K
C2 100μF
3909
+1.5V

CHIRPER

R1 15K
C1 .1μF
3909
R2 100
+1.5V
8Ω SPKR

C2 22-47μF
3909
+1.5V

SUN POWERED OSCILLATOR

R1 10K
C1 .1μF
3909
R2 100
8Ω SPKR
SILICON SOLAR CELLS

TOY ORGAN

R1 R2 R3 R4 R5
C1 1μF
3909
R6 100
+1.5V
8Ω SPKRS

R1 – 3.3k (376 Hz)
R2 – 1.0k (719 Hz)
R3 – 330 (1.73 kHz)
R4 – 150 (2.88 kHz)
R5 – 47 (6.54 kHz)

TTL CONTROLLED 3909

+5
1/6 7404
IN
R1 100
3909
C1 100μF

H = FLASH LED
L = INHIBIT

+5V
1/6 7404
IN
3909
R1 100
C1 .1μF

H = TONE + LED GLOWS
L = INHIBIT

105

DOT/BAR DISPLAY DRIVER LM3914N

ONE OF THE MOST IMPORTANT CHIPS IN THIS NOTEBOOK. LIGHTS UP TO 10 LEDs (BAR MODE) OR 1-OF-10 LEDs (DOT MODE) IN RESPONSE TO AN INPUT VOLTAGE. CHIP CONTAINS A VOLTAGE DIVIDER AND 10 COMPARATORS THAT TURN ON IN SEQUENCE AS THE INPUT VOLTAGE RISES. HERE'S A SIMPLIFIED VERSION OF THE CIRCUIT:

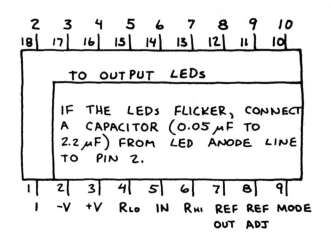

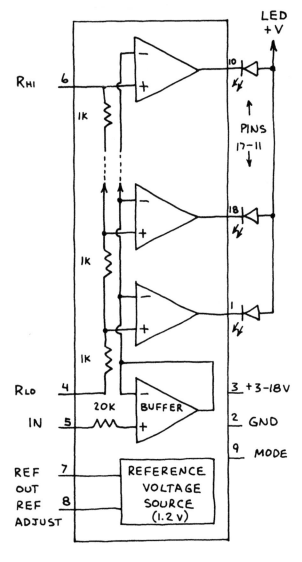

R_{HI} AND R_{LO} ARE THE ENDS OF THE DIVIDER CHAIN. THE REFERENCE VOLTAGE OUTPUT (REF OUT) IS 1.2 — 1.3 VOLTS. CONNECT PIN 9 TO PIN 11 FOR DOT MODE OR +V FOR BAR MODE.

DOT/BAR DISPLAY

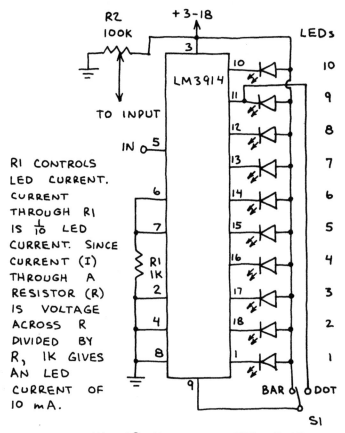

R1 CONTROLS LED CURRENT. CURRENT THROUGH R1 IS $\frac{1}{10}$ LED CURRENT. SINCE CURRENT (I) THROUGH A RESISTOR (R) IS VOLTAGE ACROSS R DIVIDED BY R, 1K GIVES AN LED CURRENT OF 10 mA.

WHEN +V = +3-18 VOLTS, THE READOUT RANGE IS 0.13 — 1.30 VOLTS. TO CHANGE RANGE TO 0.1—1.0 VOLT (0.1 VOLT PER LED), INSERT A 5K POTENTIOMETER BETWEEN PINS 6 AND 7. CONNECT VOLTMETER ACROSS PINS 5 AND 8 AND ADJUST R2 FOR 1 VOLT AT PIN 5. THEN ADJUST 1K POT UNTIL LED 10 GLOWS. REPEAT THIS PROCEDURE FOR 0.1 VOLT AT PIN 5 AND LED 1. OK TO REPLACE THE 1K POT WITH A FIXED RESISTOR OF THE PROPER VALUE.

DOT/BAR DISPLAY DRIVER (CONTINUED)
LM3914N

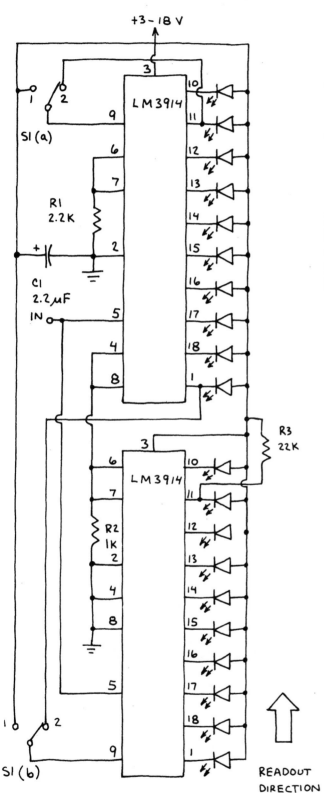

20-ELEMENT READOUT

THIS CIRCUIT SHOWS HOW TO CASCADE 2 OR MORE LM3914's. WHEN + V = 5 VOLTS, THE READOUT RANGE IS 0.14 V TO 2.7 V. HIGHEST ORDER LED STAYS ON DURING OVERRANGE. AVOID SUBSTITUTIONS FOR R1, R2 AND R3.

S1 IS THE MODE SWITCH. USE A DPDT TOGGLE. POSITION 1 SELECTS BAR AND POSITION 2 SELECTS DOT. OMIT S1 IF ONLY ONE MODE IS REQUIRED. SIMPLY WIRE IN THE CORRECT CONNECTIONS.

FLASHING BAR READOUT

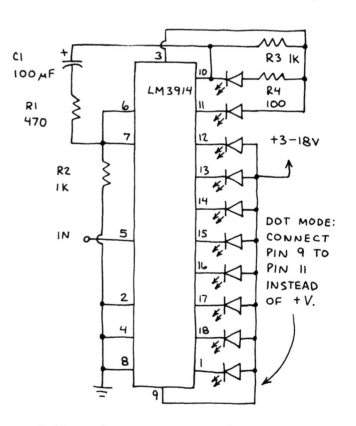

DOT MODE: CONNECT PIN 9 TO PIN 11 INSTEAD OF +V.

THE CIRCUITS ON THIS PAGE ARE ADAPTED FROM NATIONAL SEMICONDUCTOR'S LM3914 LITERATURE. BOTH WORK WELL.

WHEN ALL 10 LEDs ARE ON THE DISPLAY FLASHES. OTHERWISE THE LEDs DO NOT FLASH. INCREASE C1 TO SLOW FLASH RATE.

DOT/BAR DISPLAY DRIVER (CONTINUED)
LM3914N

SOLID-STATE OSCILLOSCOPE

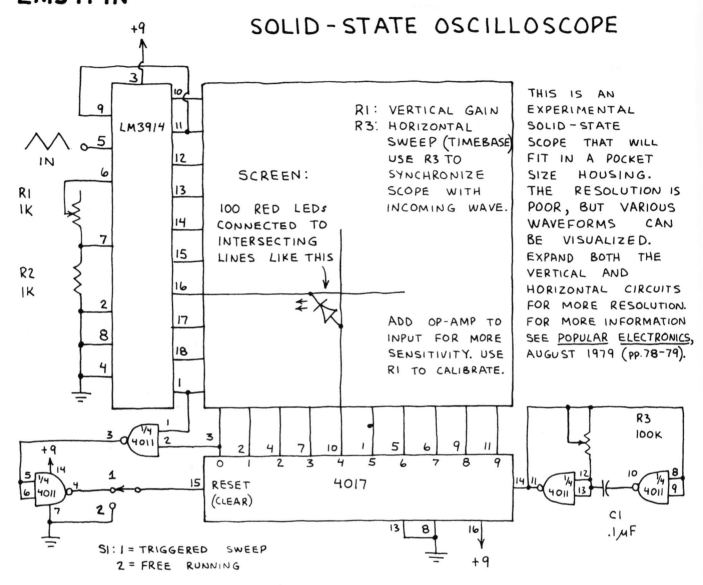

R1: VERTICAL GAIN
R3: HORIZONTAL SWEEP (TIMEBASE) USE R3 TO SYNCHRONIZE SCOPE WITH INCOMING WAVE.

SCREEN:

100 RED LEDs CONNECTED TO INTERSECTING LINES LIKE THIS

ADD OP-AMP TO INPUT FOR MORE SENSITIVITY. USE R1 TO CALIBRATE.

THIS IS AN EXPERIMENTAL SOLID-STATE SCOPE THAT WILL FIT IN A POCKET SIZE HOUSING. THE RESOLUTION IS POOR, BUT VARIOUS WAVEFORMS CAN BE VISUALIZED. EXPAND BOTH THE VERTICAL AND HORIZONTAL CIRCUITS FOR MORE RESOLUTION. FOR MORE INFORMATION SEE POPULAR ELECTRONICS, AUGUST 1979 (pp. 78-79).

S1: 1 = TRIGGERED SWEEP
2 = FREE RUNNING

USING THE LM3914 AS A CONTROLLER:

RELAY   OPTICAL COUPLING

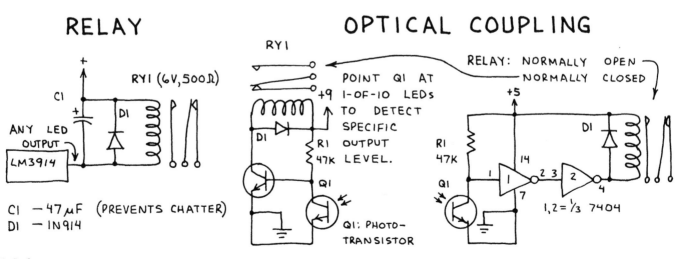

RELAY: NORMALLY OPEN
NORMALLY CLOSED

POINT Q1 AT 1-OF-10 LEDs TO DETECT SPECIFIC OUTPUT LEVEL.

Q1: PHOTO-TRANSISTOR

C1 — 47μF (PREVENTS CHATTER)
D1 — 1N914

1,2 = ⅓ 7404

DOT/BAR DISPLAY DRIVER
LM3915N

LOGARITHMIC VERSION OF THE LM3914N. THE LM3914N USES A STRING OF 1K RESISTORS AS A VOLTAGE DIVIDER WITH LINEARILY SCALED DIVISIONS. THE VOLTAGE DIVIDER RESISTORS OF THE LM3915N ARE SCALED TO GIVE A −3 dB INTERVAL FOR EACH OUTPUT. THIS CHIP IS IDEAL FOR VISUALLY MONITORING THE AMPLITUDE OF AUDIO SIGNALS.

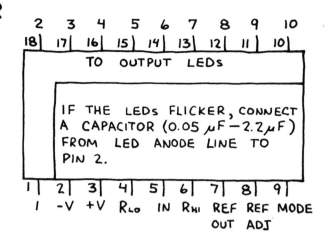

SEE LM3914N FOR EXPLANATION OF PIN FUNCTIONS.

0 TO −27 dB DOT/BAR DISPLAY

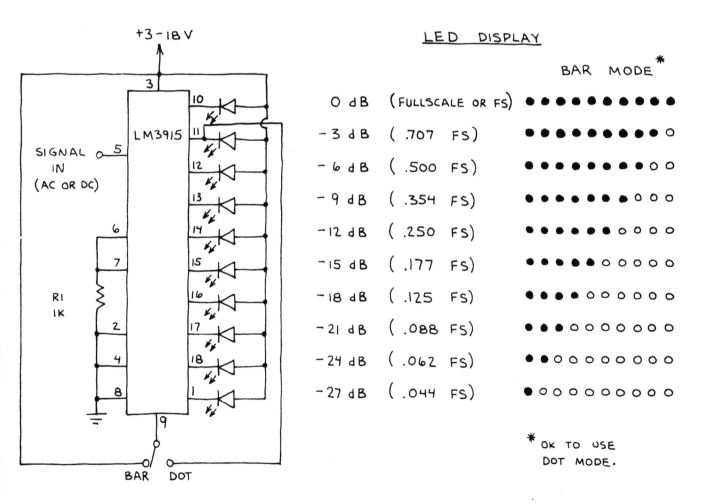

LED DISPLAY

BAR MODE *

0 dB	(FULLSCALE OR FS)	● ● ● ● ● ● ● ● ● ●
−3 dB	(.707 FS)	● ● ● ● ● ● ● ● ● ○
−6 dB	(.500 FS)	● ● ● ● ● ● ● ● ○ ○
−9 dB	(.354 FS)	● ● ● ● ● ● ● ○ ○ ○
−12 dB	(.250 FS)	● ● ● ● ● ● ○ ○ ○ ○
−15 dB	(.177 FS)	● ● ● ● ● ○ ○ ○ ○ ○
−18 dB	(.125 FS)	● ● ● ● ○ ○ ○ ○ ○ ○
−21 dB	(.088 FS)	● ● ● ○ ○ ○ ○ ○ ○ ○
−24 dB	(.062 FS)	● ● ○ ○ ○ ○ ○ ○ ○ ○
−27 dB	(.044 FS)	● ○ ○ ○ ○ ○ ○ ○ ○ ○

* OK TO USE DOT MODE.

THE INPUT SIGNAL CAN BE CONNECTED DIRECTLY TO PIN 5 WITHOUT RECTIFICATION, LIMITING OR AC COUPLING. SEE THE LM3914N FOR MORE IDEAS AND TIPS.

LED VU METER MODULE
NSM3916

INCLUDES LED BARGRAPH DRIVER
AND LEDS ON SAME SUBSTRATE.
MAKE MODE PIN HIGH FOR BAR-
GRAPH MODE. LEAVE OPEN FOR
DOT MODE. SEE DATA SUPPLIED
WITH MODULE FOR MORE INFORMA-
TION. ALSO, SEE LM3914 AND LM3915.

VU BAR GRAPH DISPLAY

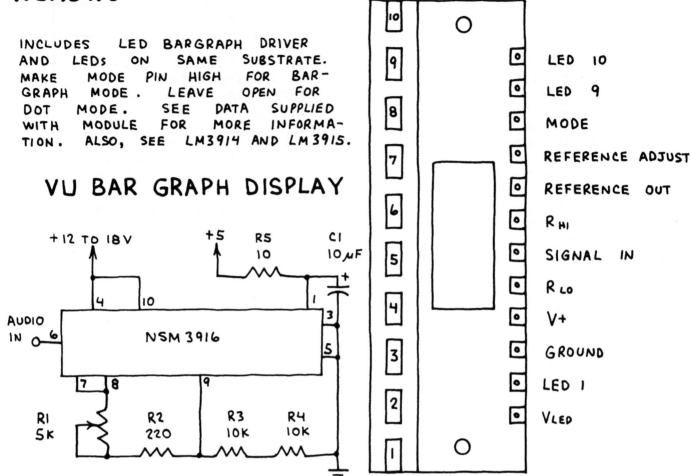

BACK AND FORTH FLASHER

RI CONTROLS
CYCLE RATE.
R4 CONTROLS
RANGE.

TWO GATE OSCILLATOR
SWITCHES 4066 ON
AND OFF. C2 IS
CHARGED VIA R2
AND DISCHARGED
BY R3 TO GIVE
VOLTAGE RAMP

1 = 1/3 4049
(GROUND UNUSED
INPUTS — PINS
7, 9, 11, 14)

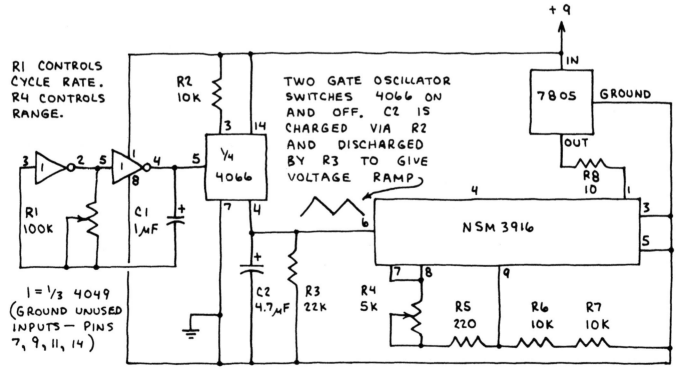

LCD CLOCK MODULE
PCIM-161

COMPLETE CLOCK MODULE.
REQUIRES ONLY 1.5 VOLT
CELL AND SWITCHES.
FOR COMPLETE INFORMATION
SEE DATA SUPPLIED WITH
MODULE. V_{DD} MUST NOT
EXCEED 1.6 VOLTS!

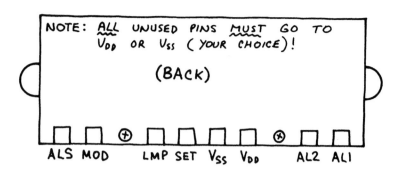

NOTE: ALL UNUSED PINS MUST GO TO V_{DD} OR V_{SS} (YOUR CHOICE)!

(BACK)

ALS MOD ⊕ LMP SET V_{SS} V_{DD} ⊗ AL2 AL1

ALARM CLOCK

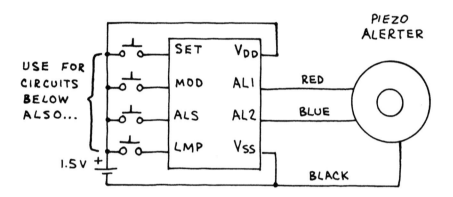

USE FOR
CIRCUITS
BELOW
ALSO...

SET V_{DD}
MOD   AL1
ALS   AL2
LMP V_{SS}

1.5V +

PIEZO ALERTER

RED
BLUE
BLACK

TO SET ALARM:

1. PRESS ALS TWICE; PRESS SET UNTIL HOUR APPEARS.

2. PRESS ALS; PRESS SET UNTIL MINUTES APPEAR.

3. PRESS ALS.

ALARM CLOCK RADIO

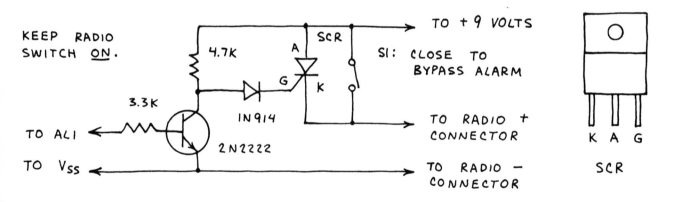

KEEP RADIO
SWITCH ON.

4.7K

3.3K

TO AL1

TO V_{SS}

2N2222

1N914

SCR
A
G   K

S1

TO +9 VOLTS

S1: CLOSE TO
BYPASS ALARM

TO RADIO +
CONNECTOR

TO RADIO −
CONNECTOR

K A G

SCR

CLOCK CONTROLLED RELAY

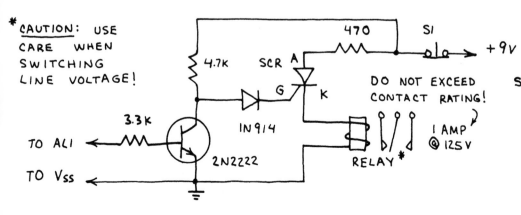

*CAUTION: USE
CARE WHEN
SWITCHING
LINE VOLTAGE!

4.7K

3.3K

TO AL1

TO V_{SS}

2N2222

1N914

SCR A
G   K

470   S1

+9V

RELAY *

DO NOT EXCEED
CONTACT RATING!

1 AMP
@ 125V

CURRENT DRAIN:
RELAY ON = 14.8 mA
RELAY OFF = 1.8 mA

S1: NORMALLY CLOSED
PUSHBUTTON.
OPEN (PRESS) TO
RESET. MUST
WAIT FOR 15
SECOND ALARM
CYCLE BEFORE
RESETTING.

TIMER 555

THE FIRST AND STILL THE MOST POPULAR IC TIMER CHIP. OPERATES AS A ONE-SHOT TIMER OR AN ASTABLE MULTIVIBRATOR. THE 556 IS TWO 555 CIRCUITS ON ONE CHIP.

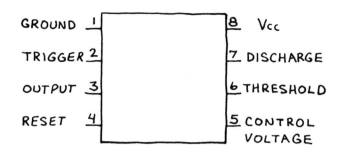

555 EQUIVALENT CIRCUIT

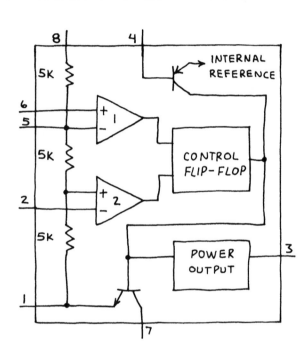

1 AND 2 ARE COMPARATORS. CIRCUIT CAN BE MADE FROM INDIVIDUAL PARTS AS SHOWN... BUT 555 IS MUCH SIMPLER.

ONE-SHOT TIMER

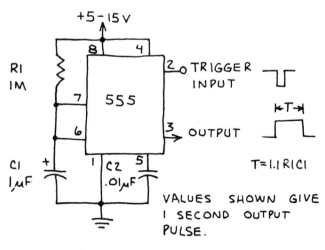

$T = 1.1 R1 C1$

VALUES SHOWN GIVE 1 SECOND OUTPUT PULSE.

BOUNCELESS SWITCH

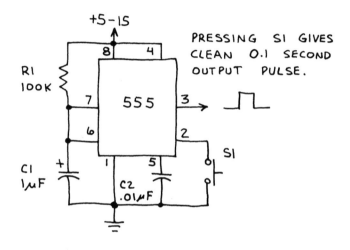

PRESSING S1 GIVES CLEAN 0.1 SECOND OUTPUT PULSE.

TIMER PLUS RELAY

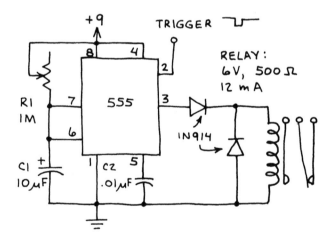

RELAY: 6V, 500 Ω 12 mA

VALUES OF R1 AND C1 SHOWN WILL PULL RELAY IN FOR UP TO ABOUT 11 SECONDS. USE POINTER KNOB AND PAPER SCALE TO HELP CALIBRATE CIRCUIT. USES INCLUDE DARKROOM TIMING. CIRCUIT CAN BE TRIGGERED BY A NEGATIVE PULSE OR WITH A PUSHBUTTON SWITCH ACROSS PINS 1 AND 2.

TIMER (CONTINUED)
555

TOY ORGAN

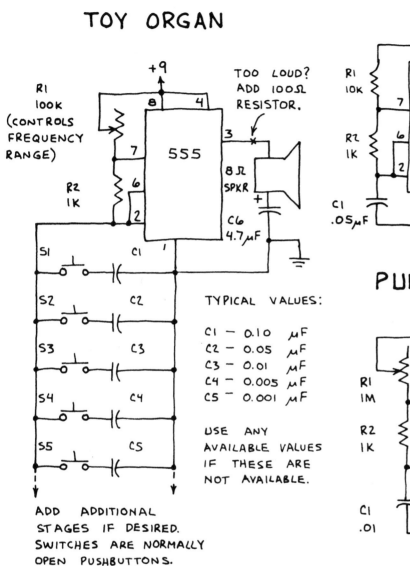

R1
100K
(CONTROLS
FREQUENCY
RANGE)

R2
1K

TOO LOUD?
ADD 100Ω
RESISTOR.

8 Ω
SPKR

C6
4.7 μF

S1 C1
S2 C2
S3 C3
S4 C4
S5 C5

TYPICAL VALUES:

C1 — 0.10 μF
C2 — 0.05 μF
C3 — 0.01 μF
C4 — 0.005 μF
C5 — 0.001 μF

USE ANY
AVAILABLE VALUES
IF THESE ARE
NOT AVAILABLE.

ADD ADDITIONAL
STAGES IF DESIRED.
SWITCHES ARE NORMALLY
OPEN PUSHBUTTONS.

LED TRANSMITTER

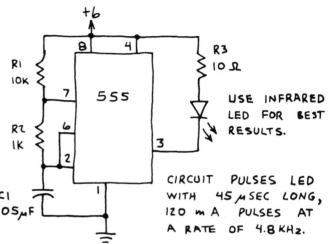

R1
10K

R2
1K

R3
10 Ω

C1
.05 μF

USE INFRARED
LED FOR BEST
RESULTS.

CIRCUIT PULSES LED
WITH 45 μSEC LONG,
120 mA PULSES AT
A RATE OF 4.8 KHz.

PULSE GENERATOR

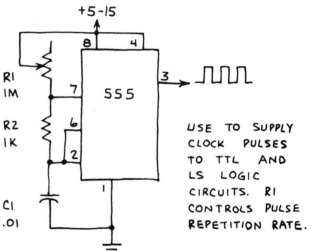

R1
1M

R2
1K

C1
.01

USE TO SUPPLY
CLOCK PULSES
TO TTL AND
LS LOGIC
CIRCUITS. R1
CONTROLS PULSE
REPETITION RATE.

MISSING PULSE DETECTOR

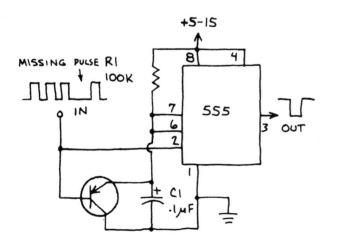

MISSING PULSE
IN

R1
100K

C1
.1 μF

3 OUT

THIS CIRCUIT IS A ONE-SHOT THAT
IS CONTINUALLY RETRIGGERED BY
INCOMING PULSES. A MISSING OR
DELAYED PULSE THAT PREVENTS
RETRIGGERING BEFORE A TIMING
CYCLE IS COMPLETE CAUSES PIN 3
TO GO LOW UNTIL A NEW INPUT
PULSE ARRIVES. R1 AND C1
CONTROL RESPONSE TIME. USE IN
SECURITY ALARMS, CONTINUITY
TESTERS, ETC.

TIMER (CONTINUED)
555

ULTRA-LONG TIME DELAY

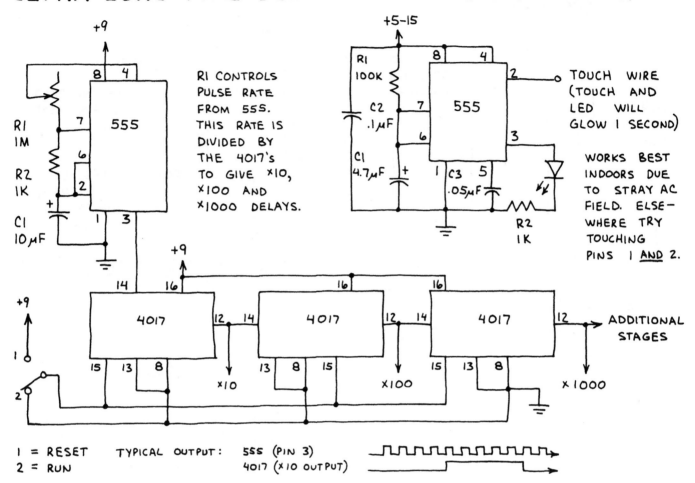

R1 CONTROLS PULSE RATE FROM 555. THIS RATE IS DIVIDED BY THE 4017's TO GIVE ×10, ×100 AND ×1000 DELAYS.

1 = RESET
2 = RUN

TYPICAL OUTPUT: 555 (PIN 3)
4017 (×10 OUTPUT)

TOUCH SWITCH

TOUCH WIRE (TOUCH AND LED WILL GLOW 1 SECOND)

WORKS BEST INDOORS DUE TO STRAY AC FIELD. ELSE-WHERE TRY TOUCHING PINS 1 AND 2.

LIGHT DETECTOR

DARK DETECTOR

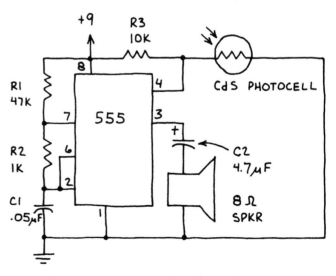

PRODUCES WARNING TONE WHEN LIGHT STRIKES PHOTOCELL. MAKES A GOOD OPEN DOOR ALARM FOR REFRIGERATOR OR FREEZER.

SILENT WHEN LIGHT STRIKES PHOTOCELL. REMOVE LIGHT AND TONE SOUNDS. FASTER RESPONSE THAN ADJACENT CIRCUIT.

TIMER (CONTINUED)
555

NEON LAMP POWER SOURCE

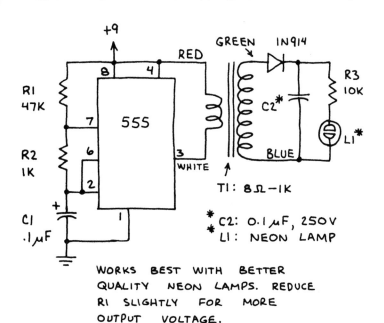

WORKS BEST WITH BETTER
QUALITY NEON LAMPS. REDUCE
R1 SLIGHTLY FOR MORE
OUTPUT VOLTAGE.

FREQUENCY DIVIDER

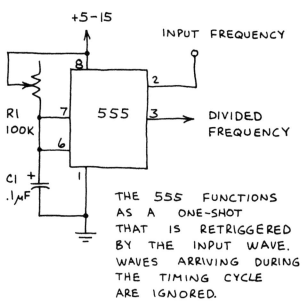

THE 555 FUNCTIONS
AS A ONE-SHOT
THAT IS RETRIGGERED
BY THE INPUT WAVE.
WAVES ARRIVING DURING
THE TIMING CYCLE
ARE IGNORED.

TRIANGLE WAVE GENERATOR

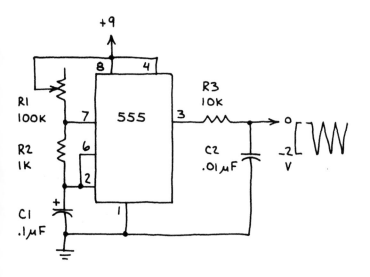

ADJUST R1 TO PROVIDE UP TO
10 KHz. OUTPUT FREQUENCY
THIS HIGH PRODUCES CLOSELY
SPACED TRIANGLE WAVES. THE
WAVES ARE SEPARATED AT SLOWER
FREQUENCIES (ᐯ—ᐯ—ᐯ).

ONE-SHOT TONE BURST

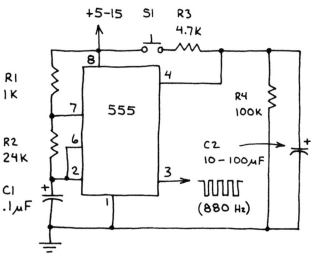

PRESS S1 AND STEADY OUTPUT
FREQUENCY APPEARS AT PIN 3.
RELEASE S1 AND OUTPUT FREQUENCY
CONTINUES UNTIL C2 IS
DISCHARGED BY R4. INCREASE
C2 (OR R4) TO INCREASE LENGTH
OF THE BURST. CHANGE FREQUENCY
OF TONE BURST VIA R2 OR C1.

DUAL TIMER
556

CONTAINS TWO INDEPENDENT
TIMERS ON A SINGLE CHIP.
BOTH TIMERS ARE IDENTICAL
TO THE 555. ALL THE
APPLICATION CIRCUITS CAN
ALSO BE BUILT WITH TWO 555's.
THIS PIN CROSS REFERENCE WILL
SIMPLIFY SUBSTITUTING TWO
555's FOR A 556 OR HALF
A 556 FOR A 555:

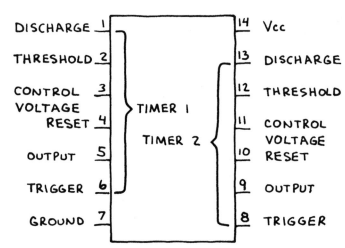

FUNCTION	555	556(1)	556(2)
GROUND	1	7	7
TRIGGER	2	6	8
OUTPUT	3	5	9
RESET	4	4	10
CONTROL V.	5	3	11
THRESHOLD	6	2	12
DISCHARGE	7	1	13
Vcc	8	14	14

INTERVAL TIMER

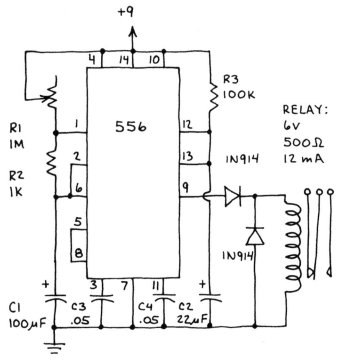

RELAY:
6V
500Ω
12 mA

TIMER 1 IS CONNECTED AS ASTABLE
OSCILLATOR. TIMER 2 IS A ONE-SHOT
RELAY DRIVER. 1 FIRES 2 ONCE EACH
CYCLE. 2 PULLS RELAY IN FOR 3-5 SECONDS.

3-STATE TONE SOURCE

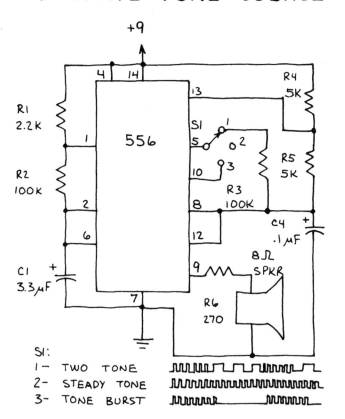

S1:
1- TWO TONE
2- STEADY TONE
3- TONE BURST

555/556 SCR OUTPUT

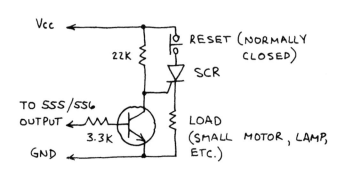

DUAL TIMER (CONTINUED)
556

SOUND SYNTHESIZER

TWO-STAGE TIMER

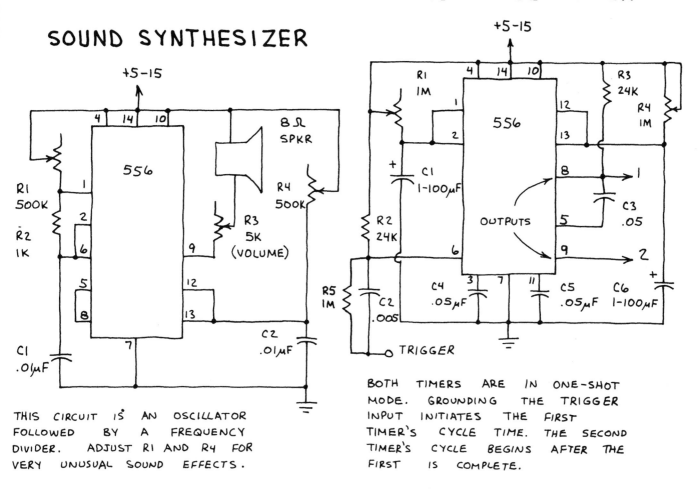

THIS CIRCUIT IS AN OSCILLATOR
FOLLOWED BY A FREQUENCY
DIVIDER. ADJUST R1 AND R4 FOR
VERY UNUSUAL SOUND EFFECTS.

BOTH TIMERS ARE IN ONE-SHOT
MODE. GROUNDING THE TRIGGER
INPUT INITIATES THE FIRST
TIMER'S CYCLE TIME. THE SECOND
TIMER'S CYCLE BEGINS AFTER THE
FIRST IS COMPLETE.

PROGRAMMABLE 4-STATE TONE GENERATOR

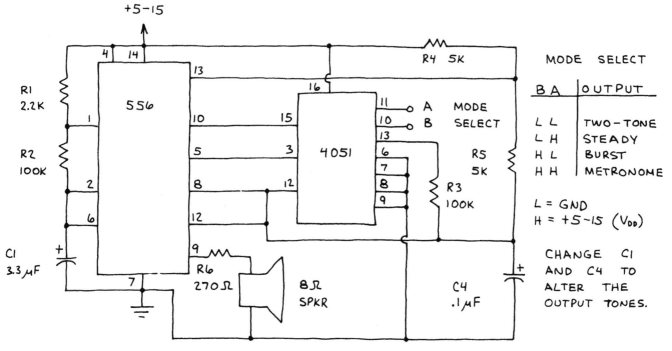

MODE SELECT

B	A	OUTPUT
L	L	TWO-TONE
L	H	STEADY
H	L	BURST
H	H	METRONOME

L = GND
H = +5-15 (V_{DD})

CHANGE C1
AND C4 TO
ALTER THE
OUTPUT TONES.

QUAD TIMER
558

CONTAINS FOUR INDEPENDENT MONOSTABLE TIMERS. EACH TIMER IS SIMILAR TO PART OF A 555 TIMER. ASTABLE OPERATION POSSIBLE WITH ONE TIMER. V_{cc} = + 4.5 TO 18 VOLTS. CONTROL AND RESET PINS ARE COMMON.

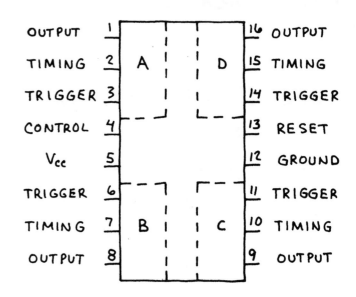

BASIC TIMER

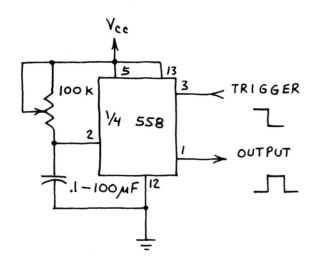

ONE-SHOT

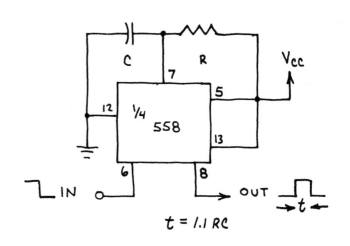

$$t = 1.1 \, RC$$

PROGRAMMABLE SEQUENCER

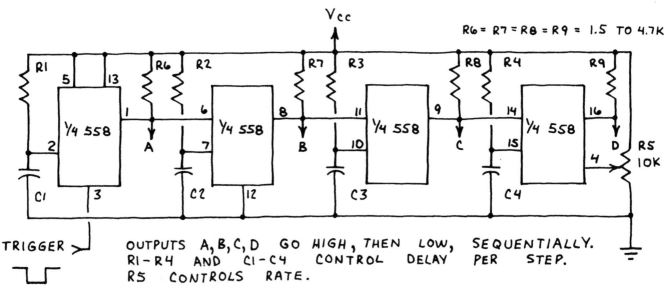

R6 = R7 = R8 = R9 = 1.5 TO 4.7K

OUTPUTS A, B, C, D GO HIGH, THEN LOW, SEQUENTIALLY. R1–R4 AND C1–C4 CONTROL DELAY PER STEP. R5 CONTROLS RATE.

FULLY ADJUSTABLE PULSE GENERATOR

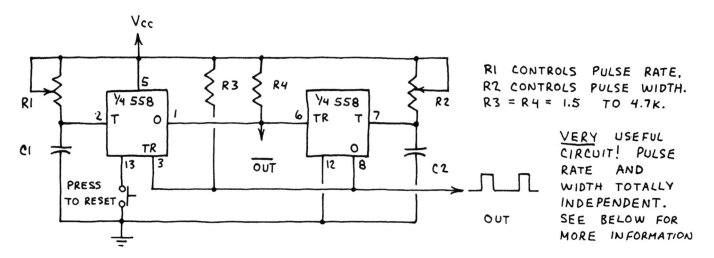

R1 CONTROLS PULSE RATE.
R2 CONTROLS PULSE WIDTH.
R3 = R4 = 1.5 TO 4.7K.

<u>VERY USEFUL CIRCUIT!</u> PULSE RATE AND WIDTH TOTALLY INDEPENDENT. SEE BELOW FOR MORE INFORMATION

SIMPLE OSCILLATOR   FIXED DUTY CYCLE PULSER

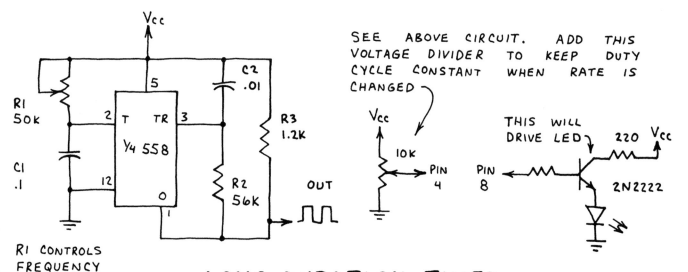

R1 CONTROLS FREQUENCY

SEE ABOVE CIRCUIT. ADD THIS VOLTAGE DIVIDER TO KEEP DUTY CYCLE CONSTANT WHEN RATE IS CHANGED

THIS WILL DRIVE LED

LONG DURATION TIMER

RS = R6 = R7 = 4.7K

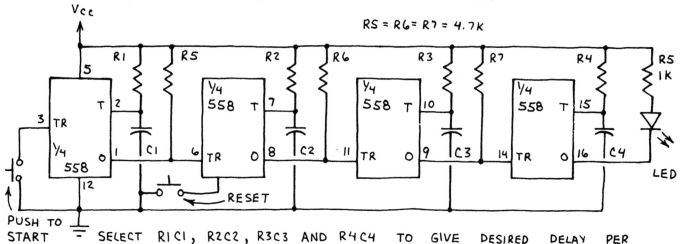

SELECT R1C1, R2C2, R3C3 AND R4C4 TO GIVE DESIRED DELAY PER STAGE. DELAY = R×C. TOTAL DELAY = SUM OF ALL STAGES. LED TURNS OFF AFTER TIME DELAY AND TURNS ON AGAIN.

119

TIMER
7555

CMOS VERSION OF THE
555. VERY LOW POWER
CONSUMPTION. WIDER
SUPPLY VOLTAGE RANGE.
LONGER TIMING CYCLES.
CAUTION: APPLY POWER TO
7555 BEFORE CONNECTING
EXTERNAL CIRCUIT.

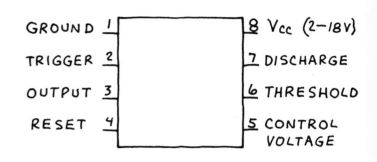

GROUND 1   8 Vcc (2-18V)
TRIGGER 2   7 DISCHARGE
OUTPUT 3   6 THRESHOLD
RESET 4   5 CONTROL VOLTAGE

FREQUENCY METER   LIGHT PROBE FOR BLIND

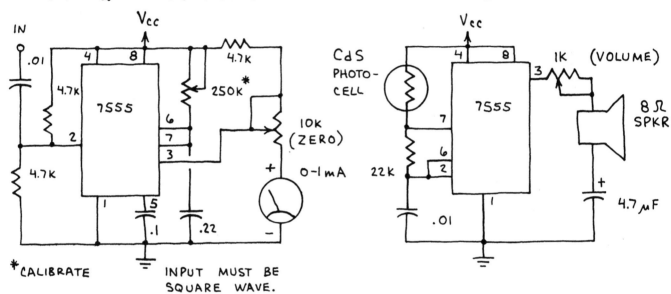

*CALIBRATE   INPUT MUST BE SQUARE WAVE.

EVENT FAILURE ALARM

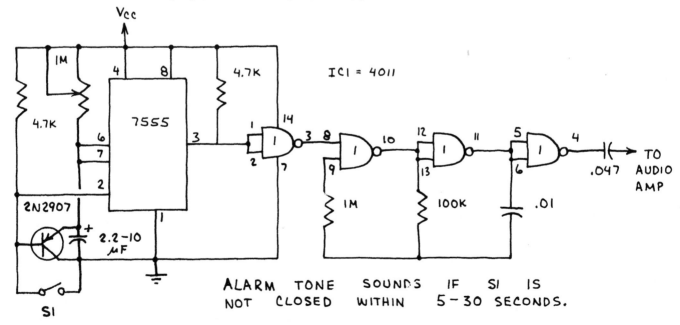

IC1 = 4011

ALARM TONE SOUNDS IF S1 IS
NOT CLOSED WITHIN 5-30 SECONDS.

PHASE-LOCKED LOOP 565

SOPHISTICATED ANALOG SYSTEM THAT AUTOMATICALLY TRACKS A FLUCTUATING INPUT SIGNAL. VOLTAGE CONTROLLED OSCILLATOR (VCO) FREQUENCY IS CONTROLLED BY OUTPUT VOLTAGE FROM PHASE COMPARATOR. THIS CAUSES VCO FREQUENCY TO MOVE TOWARD INPUT SIGNAL. THE COMPARATOR VOLTAGE OUTPUT IS AMPLIFIED AND AVAILABLE FOR COMMUNICATIONS APPLICATIONS ... AS SHOWN BELOW.

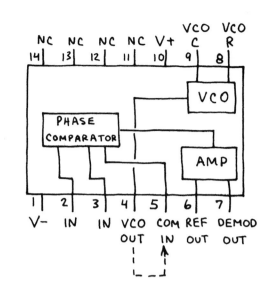

PULSE-FREQUENCY-MODULATED INFRARED COMMUNICATOR

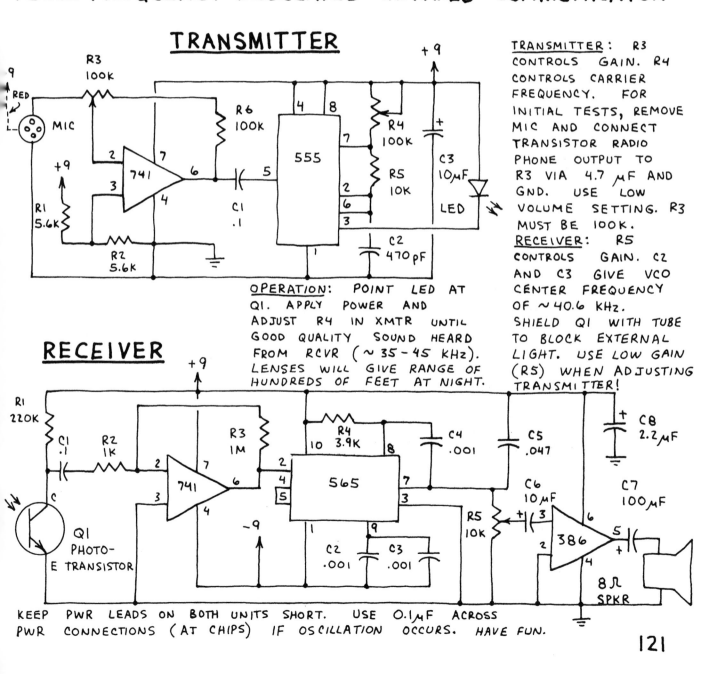

TRANSMITTER

RECEIVER

OPERATION: POINT LED AT Q1. APPLY POWER AND ADJUST R4 IN XMTR UNTIL GOOD QUALITY SOUND HEARD FROM RCVR (~ 35-45 kHz). LENSES WILL GIVE RANGE OF HUNDREDS OF FEET AT NIGHT.

TRANSMITTER: R3 CONTROLS GAIN. R4 CONTROLS CARRIER FREQUENCY. FOR INITIAL TESTS, REMOVE MIC AND CONNECT TRANSISTOR RADIO PHONE OUTPUT TO R3 VIA 4.7 μF AND GND. USE LOW VOLUME SETTING. R3 MUST BE 100K. RECEIVER: R5 CONTROLS GAIN. C2 AND C3 GIVE VCO CENTER FREQUENCY OF ~ 40.6 kHz. SHIELD Q1 WITH TUBE TO BLOCK EXTERNAL LIGHT. USE LOW GAIN (R5) WHEN ADJUSTING TRANSMITTER!

KEEP PWR LEADS ON BOTH UNITS SHORT. USE 0.1μF ACROSS PWR CONNECTIONS (AT CHIPS) IF OSCILLATION OCCURS. HAVE FUN.

121

PHASE-LOCKED LOOP (PLL) 4046

EXCEPTIONALLY VERSATILE CHIP. CONTAINS TWO PHASE COMPARATORS AND VOLTAGE CONTROLLED OSCILLATOR (VCO). USE VCO AND ONE PHASE COMPARATOR TO MAKE PLL. CIRCUITS ON <u>THIS</u> PAGE USE VCO ONLY.

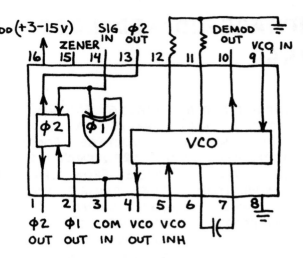

SPEAKER AMPLIFIER*

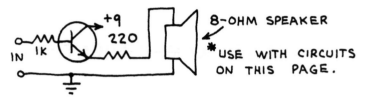

8-OHM SPEAKER

*USE WITH CIRCUITS ON THIS PAGE.

TUNABLE OSCILLATOR

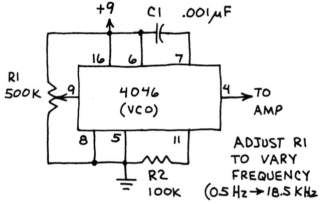

ADJUST R1 TO VARY FREQUENCY (0.5 Hz → 18.5 KHz)

CHIRP BURST SEQUENCER

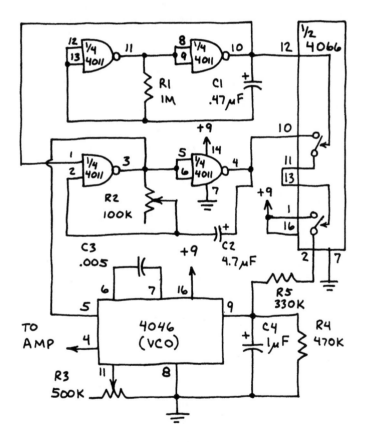

SIREN

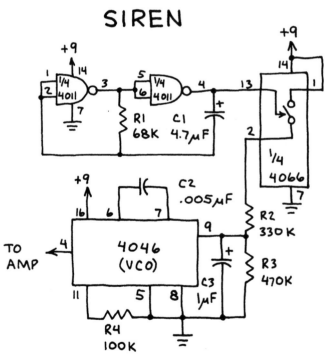

R2: ADJUST FOR 1-4 CHIRPS PER CYCLE. CHIRPS WILL HAVE DIFFERENT FREQUENCIES.

R3: CONTROLS PITCH OF CHIRPS. FOR TONES INSTEAD OF CHIRPS, CONNECT TO PIN 12 INSTEAD OF PIN 11.

CHANGE R1 OR C1 TO ALTER CYCLE TIME.
CHANGE R4 OR C2 TO ALTER FREQUENCY.
CHANGE R3 OR C3 TO ALTER WAIL.

PHASE LOCKED LOOP (CONTINUED)
4046

SOUND EFFECTS GENERATOR

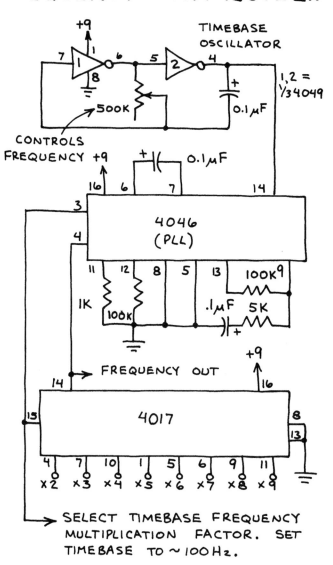

FREQUENCY SYNTHESIZER

PRODUCES FASCINATING VARIETY OF UNDULATING AND CHOPPED TONES. R1 CONTROLS CYCLE TIME. R2 CONTROLS DELAY TIME. R4 CONTROLS FREQUENCY RANGE. R5 CONTROLS CHOPPING RATE. CHANGING R5'S SETTING GIVES MOST DRAMATIC RESULTS.

SELECT TIMEBASE FREQUENCY MULTIPLICATION FACTOR. SET TIMEBASE TO ~100Hz.

LOCK INDICATOR *

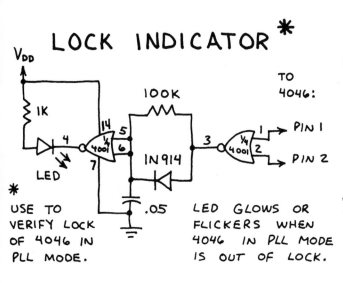

* USE TO VERIFY LOCK OF 4046 IN PLL MODE.

LED GLOWS OR FLICKERS WHEN 4046 IN PLL MODE IS OUT OF LOCK.

TONE BURST GENERATOR

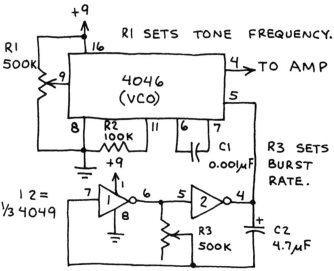

TONE DECODER
567

CONTAINS A PHASE-LOCKED LOOP. PIN 8 GOES LOW WHEN THE INPUT FREQUENCY MATCHES THE CHIP'S CENTER FREQUENCY (f_0). THE LATTER FREQUENCY IS SET BY THE TIMING RESISTOR AND CAPACITOR (R AND C) AND IS $(1.1) \div (RC)$. R SHOULD BE BETWEEN 2K-20K. THE 567 CAN BE ADJUSTED TO DETECT ANY INPUT BETWEEN 0.01 Hz TO 500kHz. NOTE: 1 SECOND OR MORE MAY BE REQUIRED FOR THE 567 TO LOCK ON TO LOW FREQUENCY INPUTS! SEE THIS CHIP'S SPECIFICATIONS FOR MORE INFORMATION.

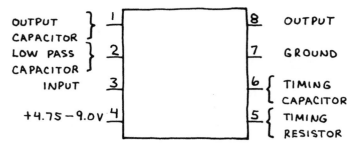

THE VALUE IN MICROFARADS OF THE LOW PASS CAPACITOR SHOULD BE n/f_0 WHERE n RANGES BETWEEN 1300 (FOR UP TO 14% f_0 DETECTION BANDWIDTH) TO 62,000 (UP TO 2% f_0 DETECTION BANDWIDTH). THE OUTPUT CAPACITOR SHOULD HAVE ABOUT TWICE THE CAPACITANCE OF THE LOW PASS FILTER CAPACITOR.

BASIC TONE DETECTOR CIRCUIT

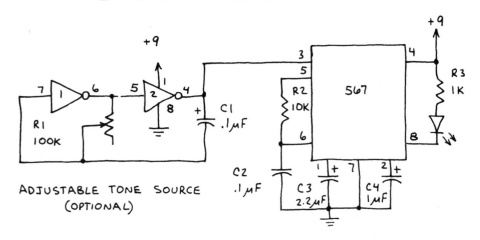

ADJUSTABLE TONE SOURCE (OPTIONAL)

THIS CIRCUIT IS HANDY FOR LEARNING TONE DECODER BASICS. THE 567 PORTION CAN BE USED IN MANY DIFFERENT APPLICATIONS (SEE BELOW). THE PREDICTED f_0 IS 1.1 KHz. THE TEST CIRCUIT f_0 WAS 1.3 KHz.

INFRARED REMOTE CONTROL SYSTEM
TRANSMITTER   RECEIVER

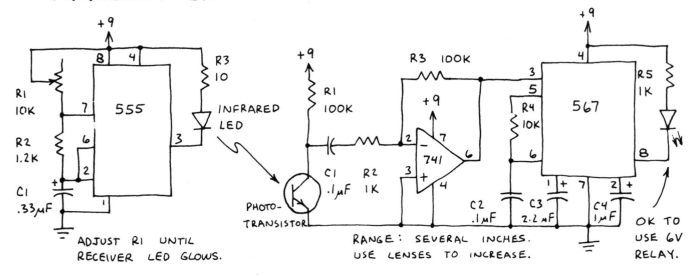

ADJUST R1 UNTIL RECEIVER LED GLOWS.

RANGE: SEVERAL INCHES. USE LENSES TO INCREASE.

OK TO USE 6V RELAY.

2-FREQUENCY OSCILLATOR

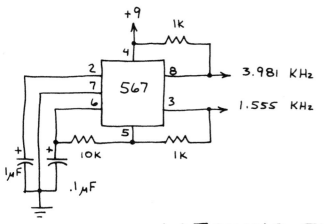

2-PHASE OSCILLATOR

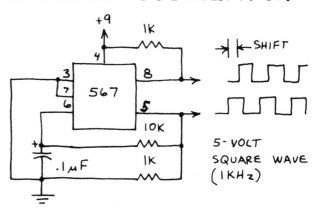

LATCHING THE 567 OUTPUT ✳

BOTH CIRCUITS SHOW ONLY THE LATCH COMPONENTS. R_L IS THE LOAD (LED, RELAY, ETC.).

✳OUTPUT STAYS ON EVEN AFTER INPUT TONE IS REMOVED.

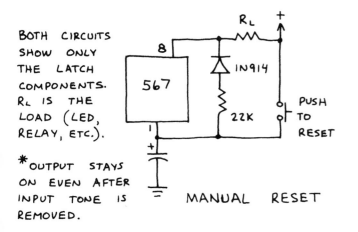

MANUAL RESET

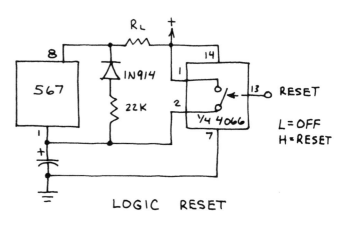

LOGIC RESET

NARROW BAND FREQUENCY DETECTOR

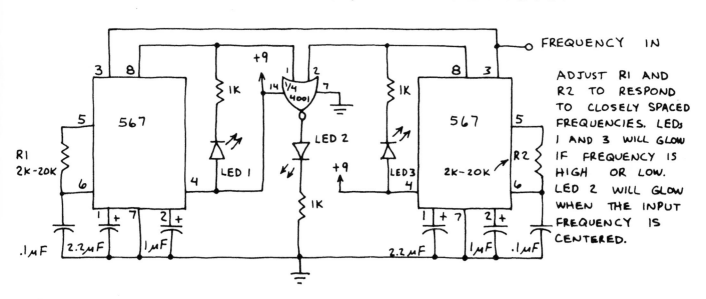

ADJUST R1 AND R2 TO RESPOND TO CLOSELY SPACED FREQUENCIES. LEDs 1 AND 3 WILL GLOW IF FREQUENCY IS HIGH OR LOW. LED 2 WILL GLOW WHEN THE INPUT FREQUENCY IS CENTERED.

TOUCH-TONE ® DECODER

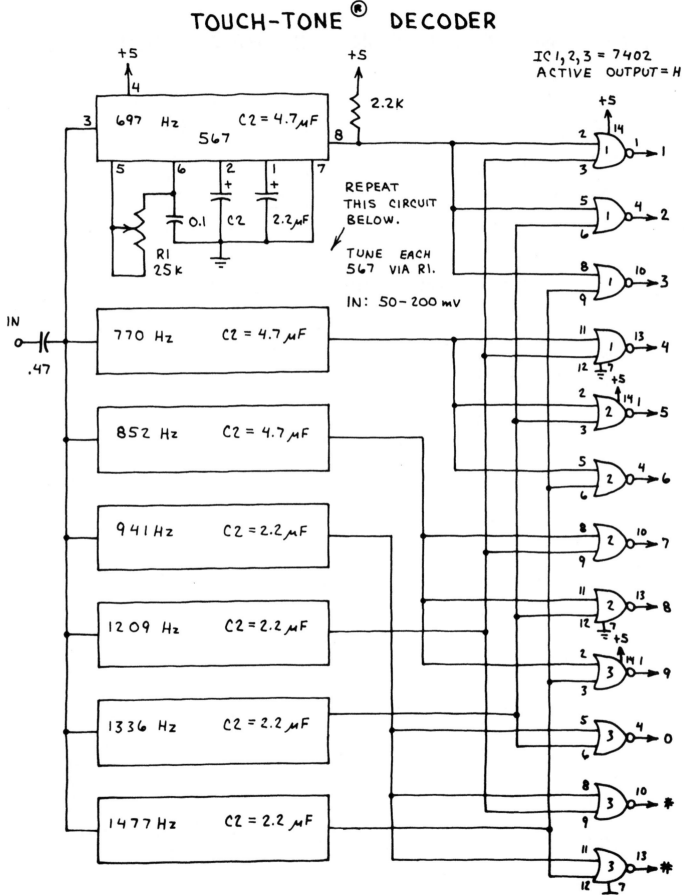

IC 1, 2, 3 = 7402
ACTIVE OUTPUT = H

REPEAT THIS CIRCUIT BELOW.

TUNE EACH 567 VIA R1.

IN: 50 – 200 mv

12-KEY PUSHBUTTON TONE MODULE
CEX-4000

GENERATES THE 12 STANDARD
TELEPHONE TONE DIALING FREQUENCY
PAIRS. V+ SHOULD <u>NOT</u> EXCEED 6
VOLTS. REQUIRES 3.58 MHz CRYSTAL.
OK TO USE FROM 1 TO 12 KEYS
FOR REMOTE CONTROL.

TOUCH-TONE® IS A REGISTERED
TRADEMARK OF AT&T.

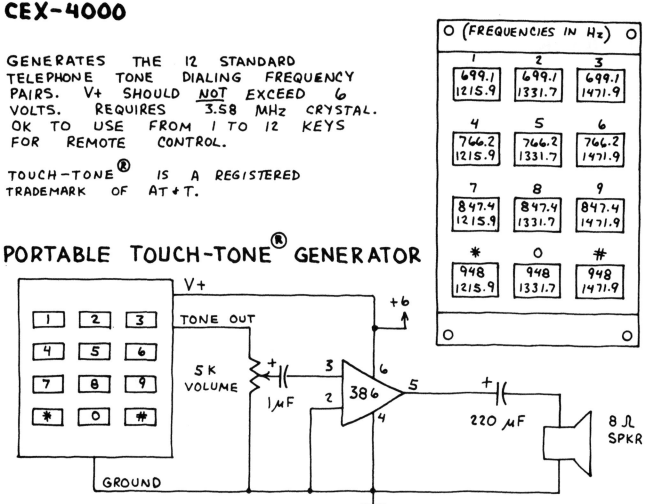

O (FREQUENCIES IN Hz) O

1	2	3
699.1 / 1215.9	699.1 / 1331.7	699.1 / 1471.9

4	5	6
766.2 / 1215.9	766.2 / 1331.7	766.2 / 1471.9

7	8	9
847.4 / 1215.9	847.4 / 1331.7	847.4 / 1471.9

*	O	#
948 / 1215.9	948 / 1331.7	948 / 1471.9

PORTABLE TOUCH-TONE® GENERATOR

REMOTE CONTROL

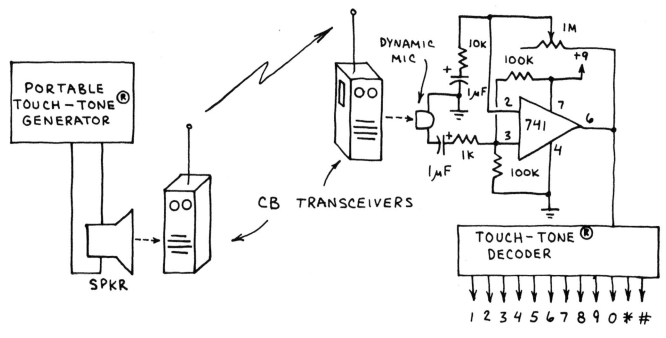

127

VOLTAGE-TO-FREQUENCY FREQUENCY-TO-VOLTAGE CONVERTER 9400

IN VOLTAGE-TO-FREQUENCY (V-F) MODE, AN INPUT VOLTAGE WHICH HAS BEEN CONVERTED INTO A CURRENT BY A RESISTOR AT PIN 3 IS TRANSFORMED INTO A PROPORTIONAL FREQUENCY. IN FREQUENCY-TO-VOLTAGE MODE A FREQUENCY AT PIN 11 IS CONVERTED INTO A PROPORTIONAL VOLTAGE. THIS CHIP CAN BE OPERATED FROM A SINGLE OR DUAL POLARITY POWER SUPPLY.

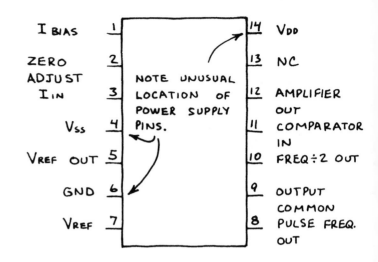

Pin		Pin	
1	I BIAS	14	V_DD
2	ZERO ADJUST	13	NC
3	I IN	12	AMPLIFIER OUT
4	Vss	11	COMPARATOR IN
5	Vref OUT	10	FREQ ÷2 OUT
6	GND	9	OUTPUT COMMON
7	Vref	8	PULSE FREQ. OUT

NOTE UNUSUAL LOCATION OF POWER SUPPLY PINS.

CAUTION: THIS CHIP INCORPORATES BOTH BIPOLAR <u>AND</u> CMOS CIRCUITRY. THEREFORE CMOS HANDLING PRECAUTIONS <u>MUST</u> BE FOLLOWED TO AVOID PERMANENT DAMAGE.

BASIC V/F CONVERTER

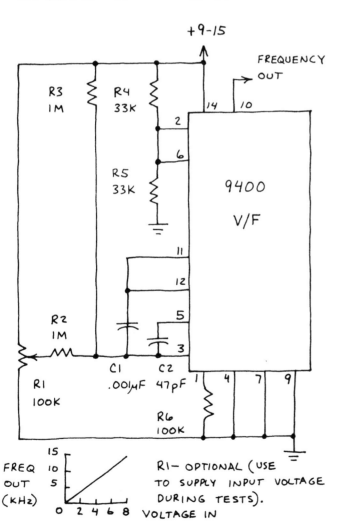

R1— OPTIONAL (USE TO SUPPLY INPUT VOLTAGE DURING TESTS).

FSK* DATA TRANSMITTER

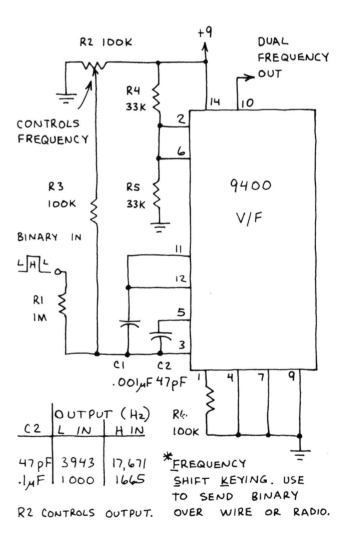

C2	OUTPUT (Hz)	
	L IN	H IN
47 pF	3943	17,671
.1μF	1000	1665

R2 CONTROLS OUTPUT.

*FREQUENCY <u>S</u>HIFT <u>K</u>EYING. USE TO SEND BINARY OVER WIRE OR RADIO.

VOLTAGE-TO-FREQUENCY (CONTINUED)
FREQUENCY-TO-VOLTAGE
CONVERTER
9400

AUDIO FREQUENCY METER

INPUT FREQUENCY MUST CROSS 0 VOLT. WORKS UP TO 25 KHz. R2 IS ZERO ADJUST FOR METER. ADJUST R7 TO GIVE MAXIMUM READING AT 25 KHz IN. FOR MORE STABILITY, CHANGE R6 TO 6-V ZENER DIODE.

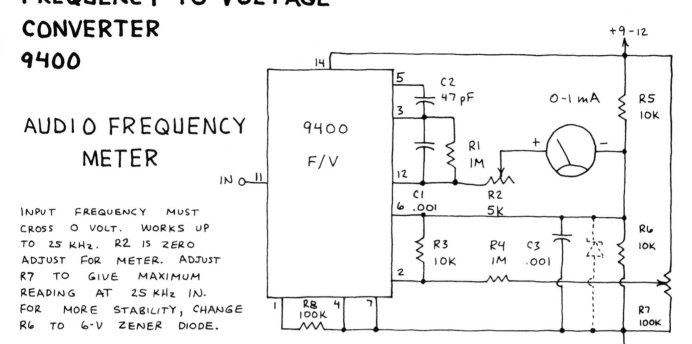

ANALOG DATA TRANSMISSION SYSTEM*

TRANSMITTER   RECEIVER

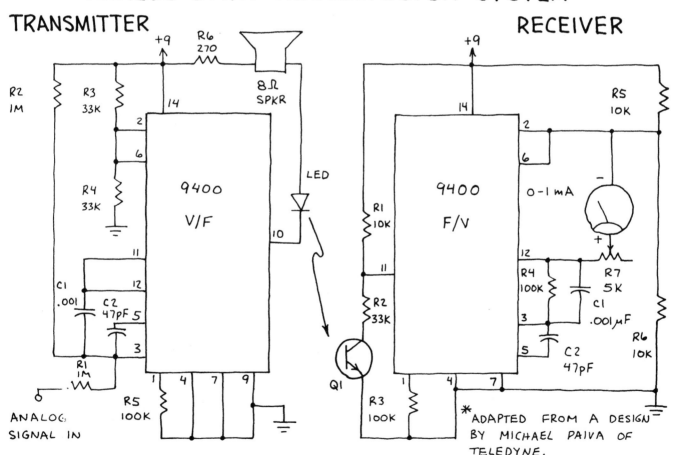

*ADAPTED FROM A DESIGN BY MICHAEL PAIVA OF TELEDYNE.

THE SPKR IS OPTIONAL BUT MAY PROVE HELPFULL DURING INITIAL TESTING. USE AN INFRARED LED. Q1 CAN BE THE PHOTOTRANSISTOR SUPPLIED WITH THE LED. R7 IN THE RECEIVER IS ZERO ADJUST.

VOLTAGE CONTROLLED OSCILLATOR (VCO) 566

VERY STABLE, EASY TO USE TRIANGLE AND SQUARE WAVE OUTPUTS. R1 AND C1 CONTROL CENTER FREQUENCY. VOLTAGE AT PIN 5 VARIES FREQUENCY. <u>IMPORTANT</u>: OUTPUT WAVE DOES NOT FALL TO 0 VOLT! AT 12 VOLTS (PIN 8), FOR EXAMPLE, TRIANGLE OUTPUT CYCLES BETWEEN +4 AND +6 VOLTS. SQUARE OUTPUT CYCLES BETWEEN +6 AND +11.5 VOLTS.

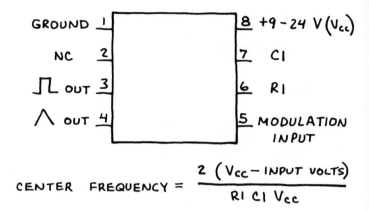

$$\text{CENTER FREQUENCY} = \frac{2\,(V_{cc} - \text{INPUT VOLTS})}{R1\,C1\,V_{cc}}$$

FUNCTION GENERATOR

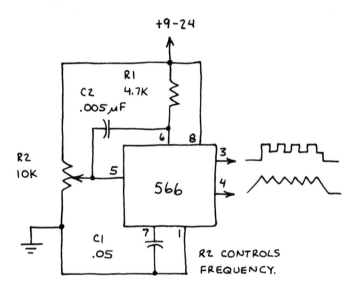

R2 CONTROLS FREQUENCY.

FSK GENERATOR *

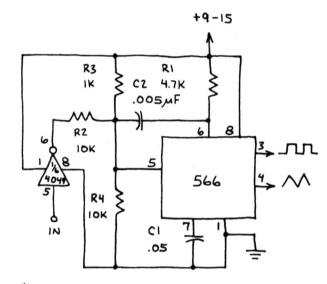

* FSK MEANS FREQUENCY SHIFT KEYING.

IN	OUTPUT
L	1.5 KHz
H	3.0 KHz

USE TO TRANSMIT BINARY DATA OVER TELEPHONE LINES OR STORE BINARY DATA ON MAGNETIC TAPE.
↳ $V_{cc} = 9$ VOLTS.

TWO-TONE WARBLER

R1 CONTROLS WARBLE RATE.

R3 CONTROLS TONE FREQUENCY.

1,2 = ⅓ 4049

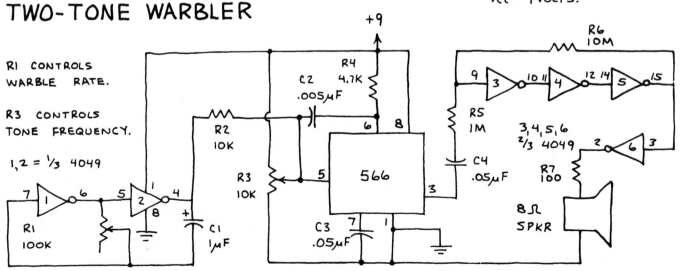

ANALOG-TO-DIGITAL CONVERTER TL507

PROVIDES ANALOG-TO-DIGITAL CONVERSION FOR MICROPROCESSORS. CAN PROVIDE 4-BIT OR 8-BIT OUTPUT WITH EXTERNAL COUNTER PLUS STEERING LOGIC. MAKES GOOD PULSE WIDTH MODULATOR. NOTE: USE $V_{cc}1$ OR $V_{cc}2$.

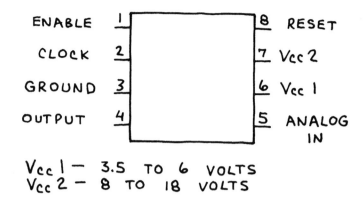

$V_{cc}1$ — 3.5 TO 6 VOLTS
$V_{cc}2$ — 8 TO 18 VOLTS

PULSE WIDTH MODULATOR

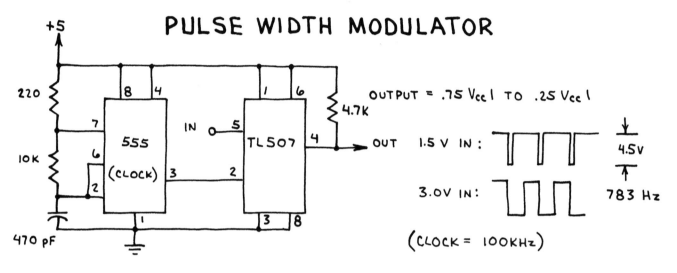

OUTPUT = .75 $V_{cc}1$ TO .25 $V_{cc}1$

1.5 V IN:

3.0V IN:

4.5V

783 Hz

(CLOCK = 100kHz)

8-BIT ANALOG-TO-DIGITAL CONVERTER

THIS PROJECT FOR ADVANCED EXPERIMENTERS.

*OMIT SECOND 74LS193 FOR 4-BIT OUTPUT. WORKS BETTER.

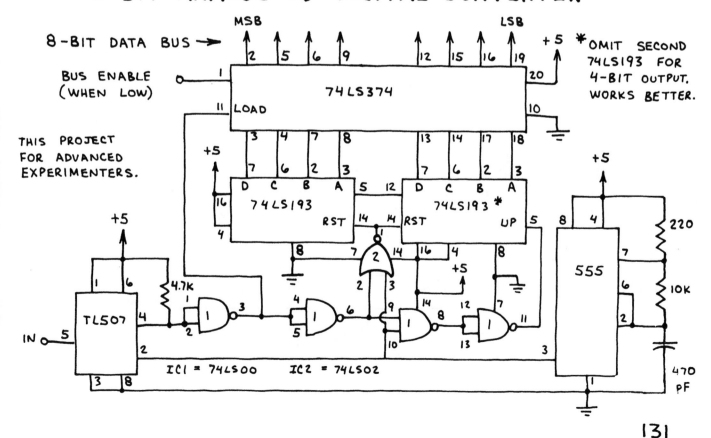

131

8-BIT DIGITAL-TO-ANALOG CONVERTER   DAC 801

PROVIDES VERY FAST 8-BIT
DIGITAL-TO-ANALOG CONVERSION.
WILL ACCEPT TTL LEVELS
AT INPUTS B1 TO B8. CAN
PROVIDE ± OUTPUT. USE
TO INTERFACE MICROCOMPUTER
TO ANALOG DEVICES.

B1 – MOST SIGNIFICANT BIT.
B8 – LEAST SIGNIFICANT BIT.
V± – ±4.5 TO 18 V.

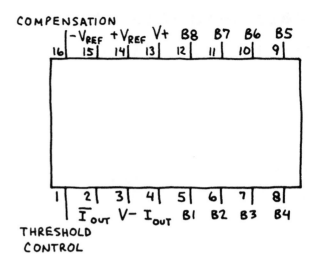

8-BIT DAC

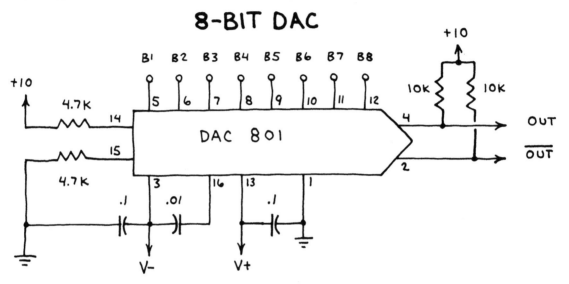

DAC 801 POWER SUPPLY

T1: 120 VAC / 25.2 VAC CT

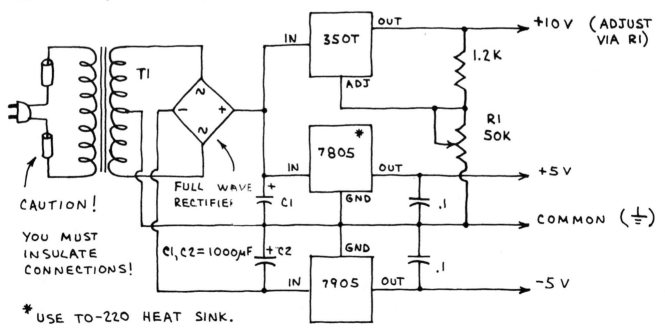

CAUTION!

YOU MUST
INSULATE
CONNECTIONS!

C1, C2 = 1000 μF

*USE TO-220 HEAT SINK.

132

8-BIT DIGITAL-TO-ANALOG CONVERTER   DAC 801   (CONTINUED)

256-STEP STAIRCASE GENERATOR

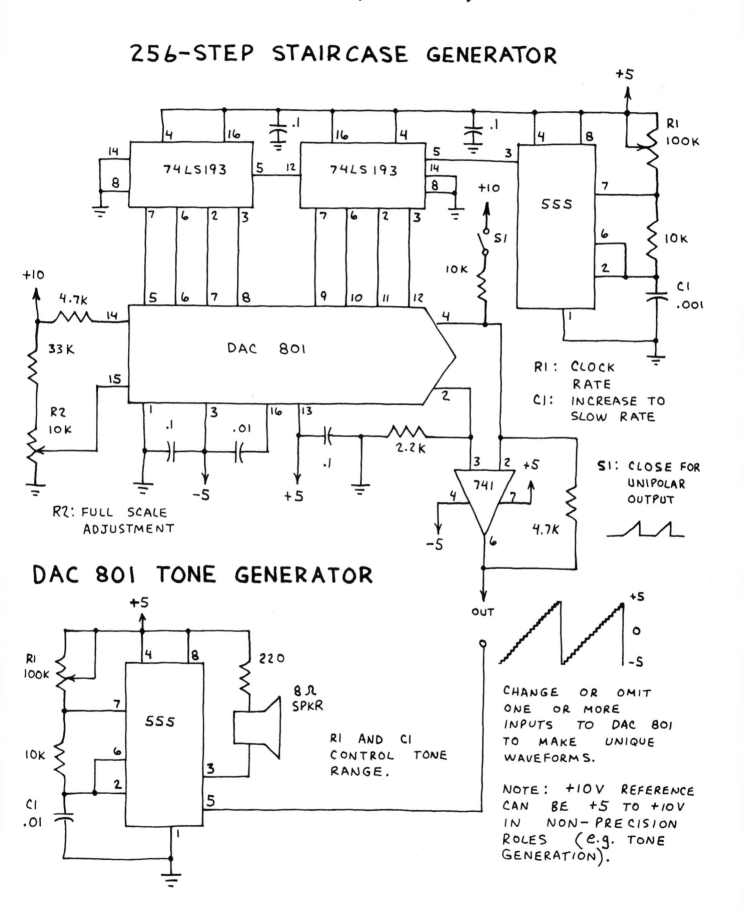

R1: CLOCK RATE
C1: INCREASE TO SLOW RATE

R2: FULL SCALE ADJUSTMENT

S1: CLOSE FOR UNIPOLAR OUTPUT

DAC 801 TONE GENERATOR

R1 AND C1 CONTROL TONE RANGE.

CHANGE OR OMIT ONE OR MORE INPUTS TO DAC 801 TO MAKE UNIQUE WAVEFORMS.

NOTE: +10V REFERENCE CAN BE +5 TO +10V IN NON-PRECISION ROLES (e.g. TONE GENERATION).

TEMPERATURE SENSOR AND ADJUSTABLE CURRENT SOURCE LM334

VERSATILE 3-LEAD COMPONENT THAT LOOKS MORE LIKE A TRANSISTOR THAN AN IC. CAN BE USED AS A TEMPERATURE SENSOR, CURRENT SOURCE FOR LEDs AND OTHER COMPONENTS OR CIRCUITS, VOLTAGE REFERENCE, ETC.

1 = R
2 = + V
3 = -V (GND)

BASIC THERMOMETERS

BASIC CURRENT SOURCE

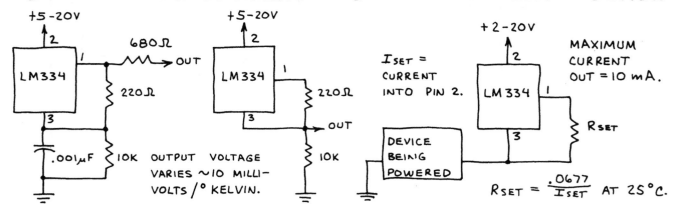

680Ω → OUT
220Ω
.001μF   10K   OUTPUT VOLTAGE VARIES ~10 MILLIVOLTS / ° KELVIN.

220Ω → OUT
10K

I_{SET} = CURRENT INTO PIN 2.

MAXIMUM CURRENT OUT = 10 mA.

R_{SET}

DEVICE BEING POWERED

$$R_{SET} = \frac{.0677}{I_{SET}} \text{ AT } 25°C.$$

VOLTAGE REFERENCE

CALIBRATED LED

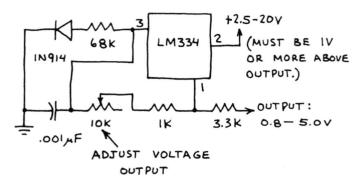

1N914   68K   LM334
+2.5-20V (MUST BE 1V OR MORE ABOVE OUTPUT.)

.001μF   10K   1K   3.3K → OUTPUT: 0.8 - 5.0V

ADJUST VOLTAGE OUTPUT

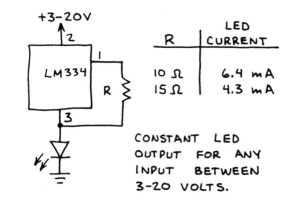

R	LED CURRENT
10 Ω	6.4 mA
15 Ω	4.3 mA

CONSTANT LED OUTPUT FOR ANY INPUT BETWEEN 3-20 VOLTS.

RAMP GENERATOR

LIGHT METER

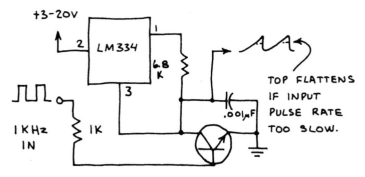

6.8 K
.001μF
1 KHz IN   1K

TOP FLATTENS IF INPUT PULSE RATE TOO SLOW.

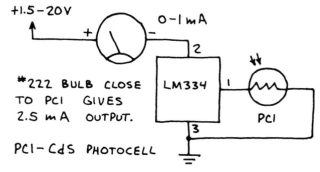

0-1 mA

#222 BULB CLOSE TO PC1 GIVES 2.5 mA OUTPUT.

PC1 - CdS PHOTOCELL

PC1

134

POWER AMPLIFIER
LM386

DESIGNED MAINLY FOR LOW
VOLTAGE AMPLIFICATION. WILL
DRIVE DIRECTLY AN 8-OHM
SPEAKER. GAIN FIXED AT 20
BUT CAN BE INCREASED TO
ANY VALUE UP TO 200.

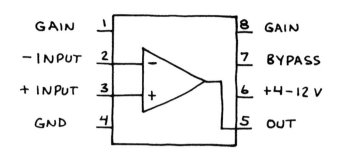

X20 AMPLIFIER

X200 AMPLIFIER

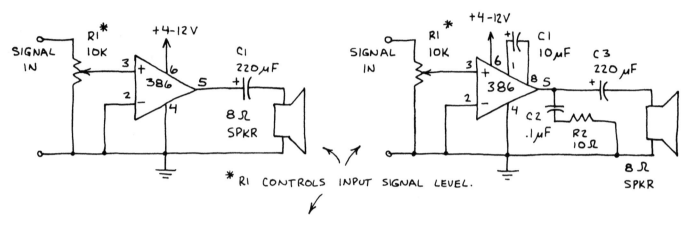

* R1 CONTROLS INPUT SIGNAL LEVEL.

BASS BOOSTER

AUDIBLE ALARM

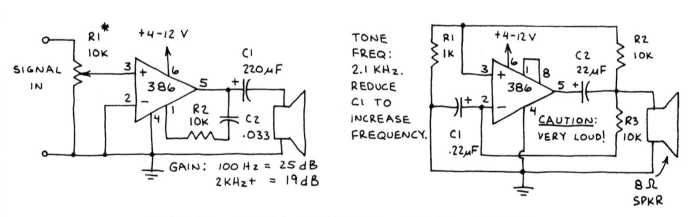

GAIN: 100 Hz = 25 dB
2 KHz+ = 19 dB

TONE
FREQ:
2.1 KHz.
REDUCE
C1 TO
INCREASE
FREQUENCY.

CAUTION:
VERY LOUD!

HIGH GAIN POWER AMPLIFIER

CIRCUIT SHOWN
IS VERY SENSITIVE
LIGHT WAVE
RECEIVER. OK TO
USE OTHER OP-
AMPS FOR THE
TL084.

Q1-PHOTOTRANSISTOR

USE CARE;
SPKR CAN
BE LOUD!

GAIN = 20 (TO CHANGE
SEE ABOVE).

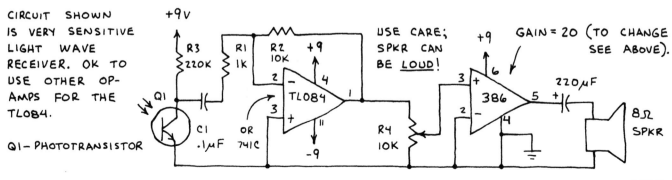

135

8-WATT POWER AMPLIFIER
LM383 / TDA2002

POWER AMPLIFIER DESIGNED SPECIFICALLY
FOR AUTOMOTIVE APPLICATIONS — BUT
IDEAL FOR ANY AUDIO AMPLIFICATION
SYSTEM. DESIGNED TO DRIVE A 4-OHM
LOAD (EQUIVALENT TO A SINGLE 4-OHM
SPEAKER OR TWO 8-OHM SPEAKERS
IN PARALLEL). THIS CHIP CONTAINS
THERMAL SHUTDOWN CIRCUITRY TO
PROTECT ITSELF FROM EXCESSIVE LOADING.
THIS WILL CAUSE SEVERE DISTORTION
DURING OVERLOAD CONDITIONS. YOU <u>MUST</u>
USE AN APPROPRIATE HEAT SINK.
                              SPREAD
SOME HEAT SINK COMPOUND
ON THE LM383 TAB BEFORE ATTACHING
THE HEAT SINK.

NOTE PRE-
FORMED LEADS.

1 - + IN
2 - − IN
3 - GND
4 - OUT
5 - +5 - 20V

1 2 3 4 5

8-WATT AMPLIFIER

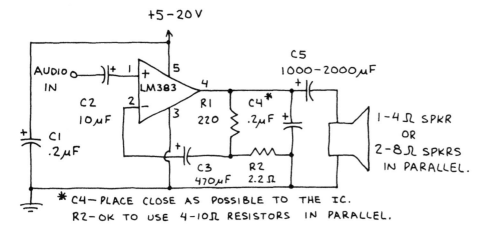

OPERATION:

1. USE HEAT SINK.
2. REDUCE POWER SUPPLY
   VOLTAGE TO 6-9 VOLTS
   (AS IN CIRCUIT BELOW)
   IF SEVERE DISTORTION
   OCCURS.
3. DON'T APPLY EXCESSIVE
   INPUT SIGNAL.

* C4—PLACE CLOSE AS POSSIBLE TO THE IC.
R2—OK TO USE 4-10Ω RESISTORS IN PARALLEL.

16-WATT BRIDGE AMPLIFIER

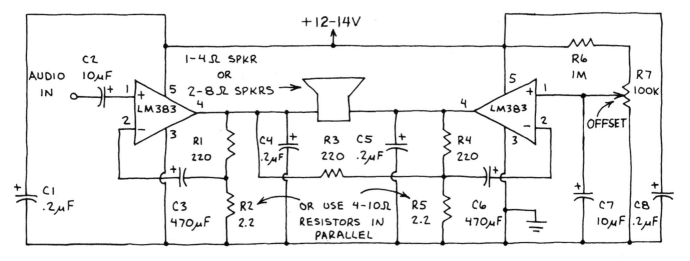

DUAL 2-WATT AMPLIFIER
LM1877/LM377

HIGH QUALITY, EASY TO USE POWER
AMPLIFIER. IDEAL FOR DO-IT-YOURSELF
STEREO, P.A. SYSTEMS, INTERCOMS, ETC.
AUTOMATIC THERMAL SHUTDOWN PROTECTS
AGAINST OVERHEATING. 70 dB CHANNEL
SEPARATION MEANS VIRTUALLY NO
CROSSTALK. ONLY 3 MICROVOLTS NOISE INPUT.
HEATSINKING: UNNECESSARY IN MANY
APPLICATIONS SINCE AVERAGE POWER IS
USUALLY WELL BELOW BRIEF PEAKS. IN
ANY CASE, PINS 3, 4, 5, 10, 11 AND 12 SHOULD
BE CONNECTED TOGETHER. IF LOAD EXCEEDS
DEVICE RATING, THERMAL SHUTDOWN WILL
OCCUR.... AND WILL CAUSE SEVERE DISTORTION.
USE HEATSINK (UP TO 10 SQUARE INCHES OF
COPPER FOIL ON PC BOARD OR METAL FIN)
IF THIS OCCURS.

NOTE: GND PINS SHOULD
BE HEAT SUNK FOR
MAXIMUM POWER.

STEREO AMPLIFIER

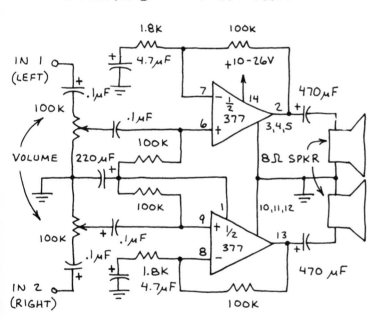

4-WATT AMPLIFIER

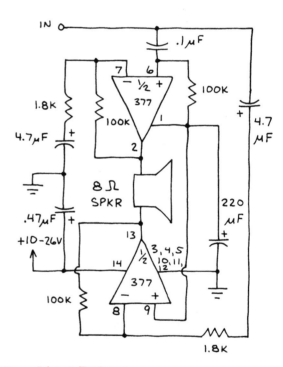

PUBLIC ADDRESS SYSTEM

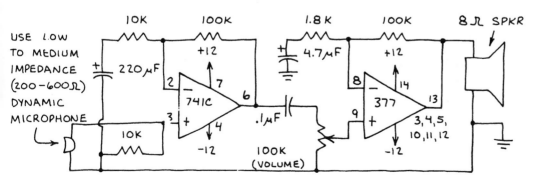

THIS CIRCUIT
WORKS WELL.
NOTE FEWER
PARTS IN
LM1877 / LM377
STAGE... THANKS
TO SPLIT POWER
SUPPLY.

USE LOW
TO MEDIUM
IMPEDANCE
(200-600Ω)
DYNAMIC
MICROPHONE

137

COMPLEX SOUND GENERATOR
SN76477N

NOTE: THE SN76488 INCLUDES BUILT-IN SPEAKER AMPLIFIER. THE SN76477 DOES NOT.

INCORPORATES S.L.F. (SUPER LOW FREQUENCY OSCILLATOR), VCO (VOLTAGE CONTROLLED OSCILLATOR), NOISE GENERATOR AND A MIXER THAT ALLOWS THE OUTPUTS FROM ONE OR MORE OF THE ABOVE TO BE COMBINED. CAN BE OPERATED TOGETHER WITH APPROPRIATE RESISTORS AND CAPACITORS TO PRODUCE MANY KINDS OF SOUNDS. CAN BE CONTROLLED BY EXTERNAL LOGIC. SEE DATA SUPPLIED WITH CHIP FOR MORE INFO.

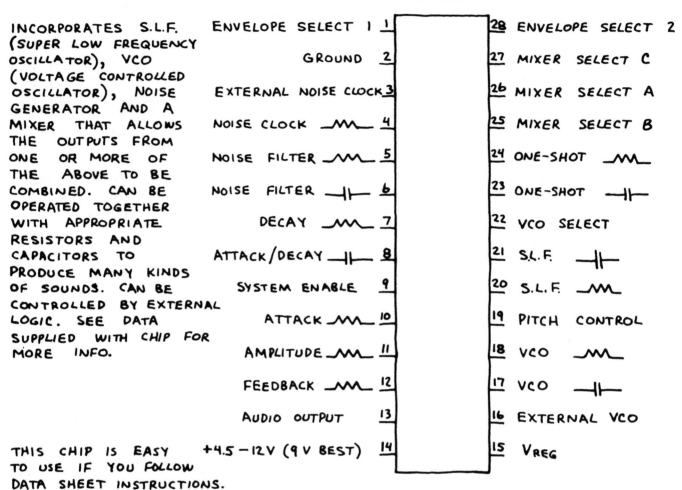

THIS CHIP IS EASY TO USE IF YOU FOLLOW DATA SHEET INSTRUCTIONS.

PERCUSSION SYNTHESIZER

S1 — PRESS TO ACTIVATE SOUND.

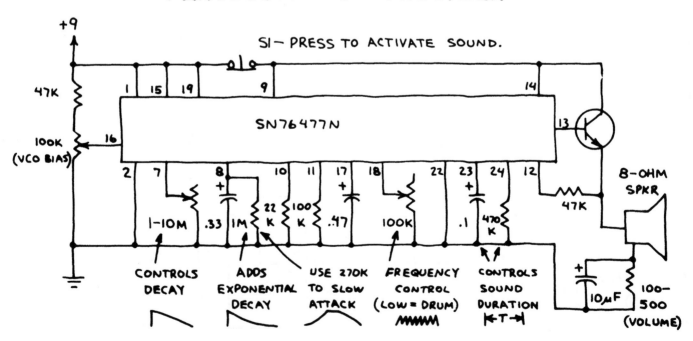

COMPLEX SOUND GENERATOR (CONTINUED)
SN76477N /

NOISE GENERATOR

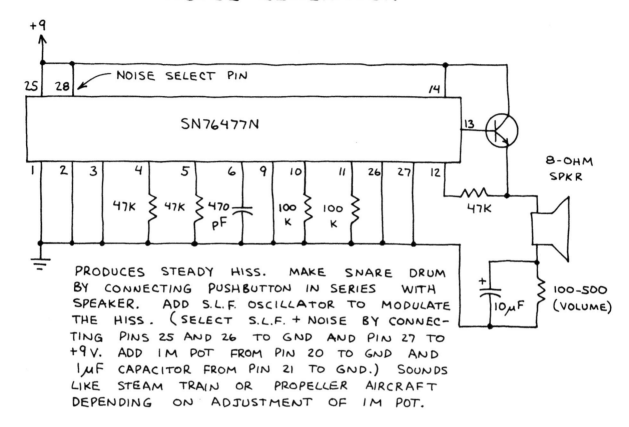

PRODUCES STEADY HISS. MAKE SNARE DRUM BY CONNECTING PUSHBUTTON IN SERIES WITH SPEAKER. ADD S.L.F. OSCILLATOR TO MODULATE THE HISS. (SELECT S.L.F. + NOISE BY CONNECTING PINS 25 AND 26 TO GND AND PIN 27 TO +9V. ADD 1M POT FROM PIN 20 TO GND AND 1μF CAPACITOR FROM PIN 21 TO GND.) SOUNDS LIKE STEAM TRAIN OR PROPELLER AIRCRAFT DEPENDING ON ADJUSTMENT OF 1M POT.

UNIVERSAL UP-DOWN TONE GENERATOR

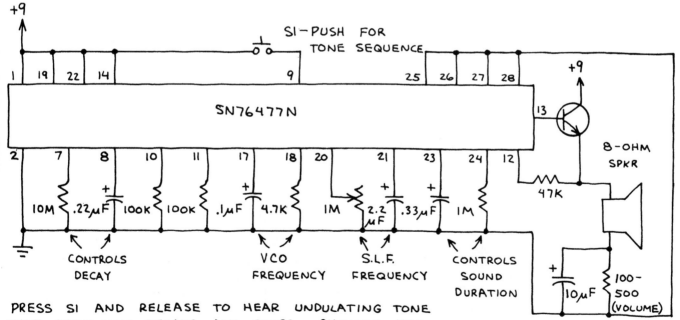

PRESS S1 AND RELEASE TO HEAR UNDULATING TONE THAT GRADUALLY DECAYS AND STOPS. CHANGE VCO AND S.L.F. COMPONENTS FOR MANY DIFFERENT SOUND EFFECTS RANGING FROM SIREN TO SCIENCE FICTION MOVIE SOUNDS. FOR CONTINUOUS SOUND, OMIT COMPONENTS AT PINS 7, 8, 23, 24 AND GROUND PIN 9.

COMPLEX SOUND GENERATOR
SN76488N

MODIFIED VERSION OF SN76477N. INCLUDES BUILT-IN AMPLIFIER FOR DIRECT SPEAKER DRIVE. NOTE THAT SN76488N AND SN76477N HAVE DIFFERENT PINOUTS.

MANY DIFFERENT SOUNDS CAN BE CREATED. FOR BEST RESULTS, STUDY CAREFULLY THE TECHNICAL DATA SUPPLIED WITH CHIP.

VERY EASY TO DEVISE YOUR OWN UNIQUE SOUNDS!

NOTE: SOUND OUTPUT MAY CHANGE AS Vcc GOES FROM +6 TO +9V.

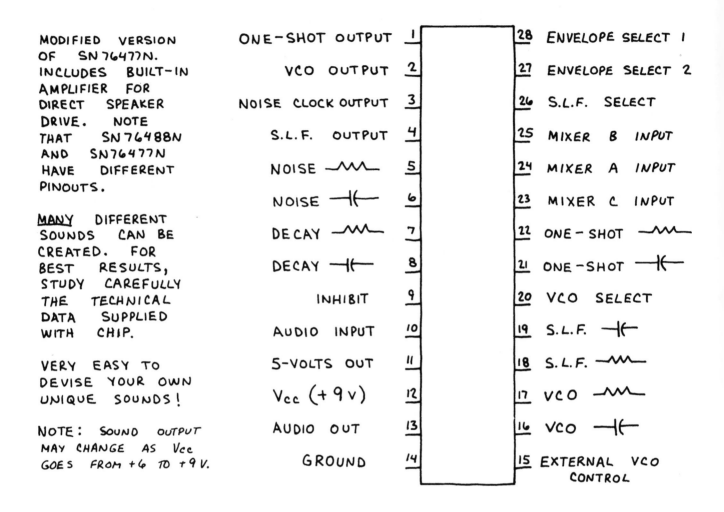

Pin	Left		Pin	Right
1	ONE-SHOT OUTPUT		28	ENVELOPE SELECT 1
2	VCO OUTPUT		27	ENVELOPE SELECT 2
3	NOISE CLOCK OUTPUT		26	S.L.F. SELECT
4	S.L.F. OUTPUT		25	MIXER B INPUT
5	NOISE ⌁		24	MIXER A INPUT
6	NOISE ⊣⊢		23	MIXER C INPUT
7	DECAY ⌁		22	ONE-SHOT ⌁
8	DECAY ⊣⊢		21	ONE-SHOT ⊣⊢
9	INHIBIT		20	VCO SELECT
10	AUDIO INPUT		19	S.L.F. ⊣⊢
11	5-VOLTS OUT		18	S.L.F. ⌁
12	Vcc (+9v)		17	VCO ⌁
13	AUDIO OUT		16	VCO ⊣⊢
14	GROUND		15	EXTERNAL VCO CONTROL

BOMB DROP PLUS EXPLOSION

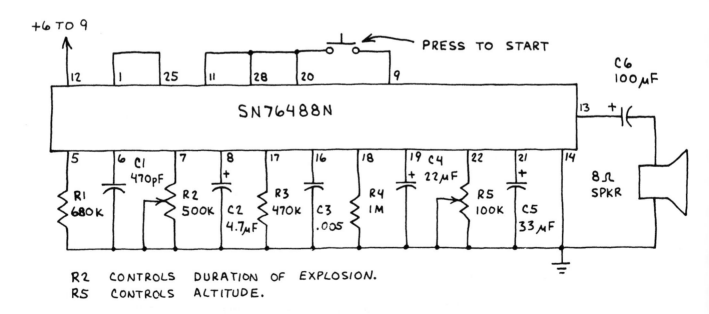

R2 CONTROLS DURATION OF EXPLOSION.
R5 CONTROLS ALTITUDE.

COMPLEX SOUND GENERATOR (CONTINUED)
SN76488N

IMPROVED STEAM ENGINE AND WHISTLE

R2 CONTROLS ENGINE SPEED.
R4 CONTROLS WHISTLE FREQUENCY.

*USE .0047 FOR RASPY WHISTLE OR .01 FOR PURE TONE.

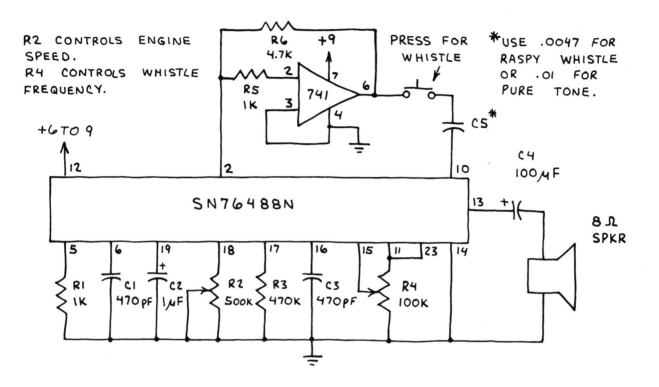

THE ULTIMATE SIREN

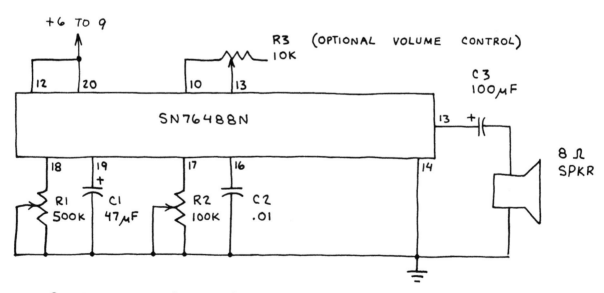

R1 CONTROLS CYCLE RATE.
R2 CONTROLS FREQUENCY.

ADJUST R1 FOR HIGH RESISTANCE TO GIVE ULTRA SLOW SIREN.

RHYTHM PATTERN GENERATOR MM5871

PRODUCES SIX DIFFERENT RHYTHM
PATTERNS AND TRIGGERS FIVE
DIFFERENT INSTRUMENTS.
ADJUSTABLE TEMPO. COMPLICATED
TO USE, BUT WELL WORTH THE EFFORT.

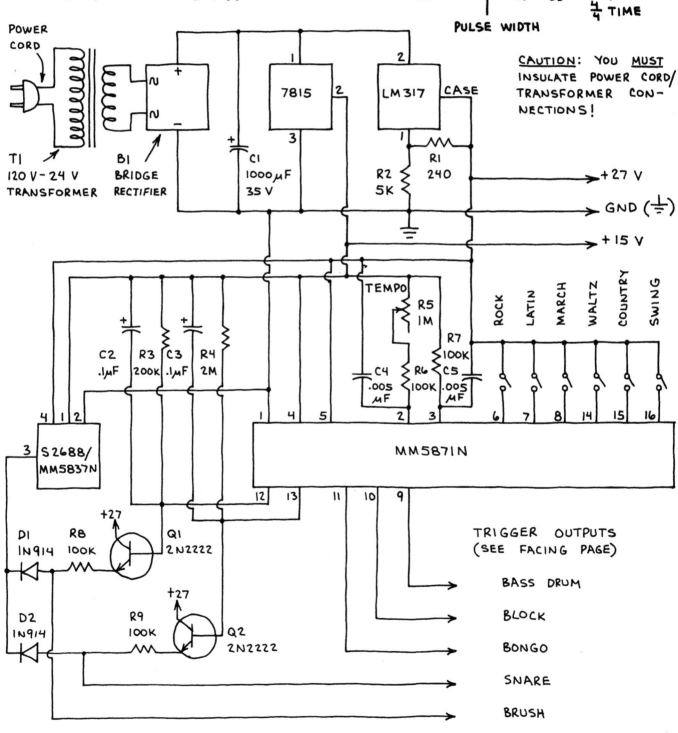

$\frac{3}{4}$ TIME   TRIGGER OUTPUTS

| 16 | 15 | 14 | 13 | 12 | 11 | 10 | 9 |

SEE FACING PAGE
FOR PIN EXPLANATIONS.

| 1 | 2 | 3 | 4 | 5 | 6 | 7 | 8 |

V_{GG} TEMPO V_{DD} V_{SS}

$\frac{4}{4}$ TIME

PULSE WIDTH

RHYTHM BOX

CAUTION: YOU **MUST**
INSULATE POWER CORD/
TRANSFORMER CON-
NECTIONS!

POWER
CORD

7815

LM317   CASE

T1
120 V - 24 V
TRANSFORMER

B1
BRIDGE
RECTIFIER

C1
1000 μF
35 V

R2
5K

R1
240

+27 V

GND ($\perp$)

+15 V

TEMPO

R5
1M

R7
100K

ROCK LATIN MARCH WALTZ COUNTRY SWING

C2
.1μF

R3
200K

C3
.1μF

R4
2M

C4
.005
μF

R6
100K

C5
.005
μF

S 2688/
MM5837N

MM5871N

D1
1N914

R8
100K

+27

Q1
2N2222

D2
1N914

R9
100K

Q2
2N2222

TRIGGER OUTPUTS
(SEE FACING PAGE)

BASS DRUM

BLOCK

BONGO

SNARE

BRUSH

RHYTHM PATTERN GENERATOR (CONTINUED)
MM5871

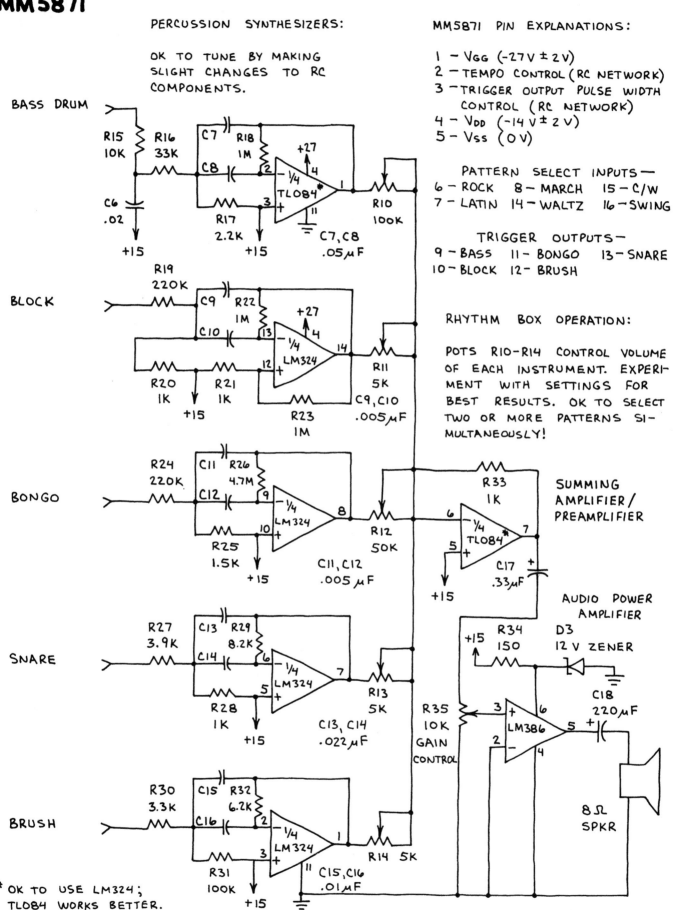

PERCUSSION SYNTHESIZERS:

OK TO TUNE BY MAKING SLIGHT CHANGES TO RC COMPONENTS.

MM5871 PIN EXPLANATIONS:

1 — V_{GG} (-27V ± 2V)
2 — TEMPO CONTROL (RC NETWORK)
3 — TRIGGER OUTPUT PULSE WIDTH CONTROL (RC NETWORK)
4 — V_{DD} (-14V ± 2V)
5 — V_{SS} (0V)

PATTERN SELECT INPUTS —
6 — ROCK   8 — MARCH   15 — C/W
7 — LATIN   14 — WALTZ   16 — SWING

TRIGGER OUTPUTS —
9 — BASS   11 — BONGO   13 — SNARE
10 — BLOCK 12 — BRUSH

RHYTHM BOX OPERATION:

POTS R10-R14 CONTROL VOLUME OF EACH INSTRUMENT. EXPERIMENT WITH SETTINGS FOR BEST RESULTS. OK TO SELECT TWO OR MORE PATTERNS SIMULTANEOUSLY!

SUMMING AMPLIFIER/PREAMPLIFIER

AUDIO POWER AMPLIFIER

BASS DRUM

R15 10K
R16 33K
C6 .02
+15
C7
C8
R18 1M
R17 2.2K
+27
1/4 TL084*
C7,C8 .05µF
R10 100K
+15

BLOCK

R19 220K
C9
C10
R22 1M
R20 1K
R21 1K
+15
+27
1/4 LM324
R23 1M
C9,C10 .005µF
R11 5K

BONGO

R24 220K
C11
C12
R26 4.7M
R25 1.5K
+15
1/4 LM324
C11,C12 .005µF
R12 50K
R33 1K
6
5
1/4 TL084*
7
C17 .33µF
+15

SNARE

R27 3.9K
C13
C14
R29 8.2K
R28 1K
+15
1/4 LM324
C13,C14 .022µF
R13 5K

R34 150
D3 12V ZENER
C18 220µF
+15

R35 10K GAIN CONTROL

BRUSH

R30 3.3K
C15
C16
R32 6.2K
R31 100K
+15
1/4 LM324
C15,C16 .01µF
R14 5K

3
2
LM386
6
4
5
8Ω SPKR

* OK TO USE LM324;
TL084 WORKS BETTER.

143

DUAL ANALOG DELAY LINE
SAD-1024A

CONTAINS TWO INDEPENDENT 512 STAGE
SERIAL ANALOG DELAY (SAD) LINES
(ALSO CALLED ANALOG SHIFT REGISTERS).
OK TO USE EACH 512 STAGE SAD
SEPARATELY OR IN SERIES. ANALOG
DELAYS OF UP TO ½ SECOND CAN BE
ACHIEVED. A 2-PHASE CLOCK IS REQUIRED
TO DRIVE INPUTS $\phi 1$ AND $\phi 2$. INPUT
DATA RIDES THROUGH THE SAD ON
ALTERNATING CLOCK PULSES AND
APPEAR AT THE TWO OUTPUTS AFTER
PASSING THROUGH ALL 512 STAGES.
CONNECT V_{bb} TO V_{DD} (PIN 7) OR, FOR
OPTIMUM RESULTS, TO 1 VOLT BELOW
V_{DD}. THIS CHIP CAN BE TRICKY TO
USE SINCE SEVERAL EXTERNAL
ADJUSTMENTS ARE REQUIRED. CIRCUITS
ON THIS PAGE EXPLAIN OPERATING
REQUIREMENTS WHILE A COMPLETE
CIRCUIT IS SHOWN ON FACING PAGE.

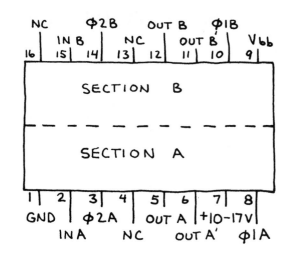

CAUTION: THIS NMOS CHIP IS
VULNERABLE TO DAMAGE FROM
STATIC DISCHARGE! FOLLOW
CMOS HANDLING PROCEDURES.

SAD IN/OUT CONTROLS

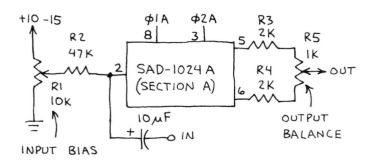

ADJUST R1 (INPUT BIAS) FOR OPTIMUM
AUDIO OUTPUT. OUTPUTS APPEAR LIKE
THIS ON A SCOPE:

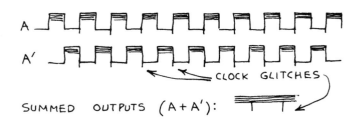

SET SCOPE TO VISUALIZE INPUT SIGNAL
(COMPRESSING CLOCK RATE):

SERIAL OPERATION

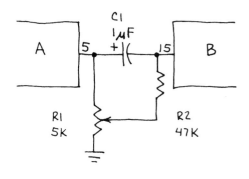

R1 CONTROLS BIAS TO SECTION B.
NOTE THAT ONLY ONE OUTPUT OF A
IS CONNECTED TO INPUT OF B.

OUTPUT SUMMER

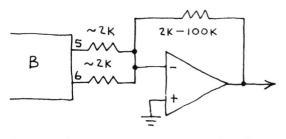

ANY OP-AMP CAN BE USED, BUT
LOW NOISE FET INPUT TYPES ARE
BEST.

DUAL ANALOG DELAY LINE (CONTINUED)
SAD-1024A

ADJUSTABLE FLANGER OR PHASER

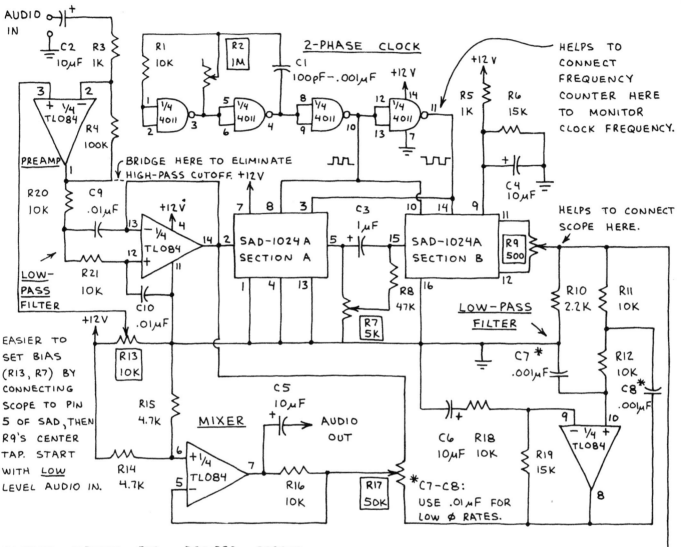

ADJUST CIRCUIT FOR DESIRED EFFECT BY CONNECTING TRANSISTOR RADIO TO AUDIO INPUT. TUNE RADIO TO A TALK SHOW FOR BEST RESULTS. R13 AND R7 CONTROL BIAS TO SECTIONS A AND B OF THE SAD. R9 BALANCES THE SAD OUTPUTS. R2 CONTROLS THE CLOCK RATE. R17 IS THE MAIN BALANCE CONTROL. IT CONTROLS THE RELATIVE AMPLITUDES OF THE ORIGINAL AND DELAYED SIGNAL APPLIED TO THE MIXER. CONNECT THE OUTPUT TO A POWER AMPLIFIER. YOU MUST ADJUST BIAS CONTROLS PROPERLY FOR BEST RESULTS. SET R2 FOR LOW FREQUENCIES (3-8KHz) FOR SINGLE ECHO. USE HIGHER CLOCK FREQUENCIES (20-100KHz) FOR HOLLOW, SWISHY SOUNDS. NOTE: THIS CIRCUIT IS NOT FOR BEGINNERS.

REVERBERATOR

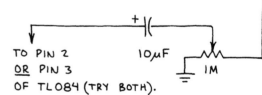

TO PIN 2 _OR_ PIN 3 OF TL084 (TRY BOTH).

ADD THIS FEEDBACK CIRCUIT FOR UNUSUAL REVERBERATION EFFECTS. SLOW CLOCK FREQUENCIES GIVE MOST STRIKING REVERBERATIONS. TRY 5-20 KHz. FASTER CLOCK (20-100 KHz) AND CAREFUL ADJUSTMENT GIVES ROBOT-LIKE SOUND USED IN SOME SCIENCE FICTION MOVIES.

TOP OCTAVE SYNTHESIZER
S50240

THIS PMOS CHIP ACCEPTS AN INPUT FREQUENCY (ϕ) AND THEN DIVIDES IT INTO A FULL OCTAVE PLUS ONE NOTE ON THE EQUALLY TEMPERED SCALE. THIS CHIP IS IDEAL FOR MUSIC SYNTHESIZERS, ORGANS, ETC. FOR TOP OCTAVE OPERATION, ϕ SHOULD BE 2.00024 MHz; LOWER FREQUENCIES GIVE LOWER OCTAVES.

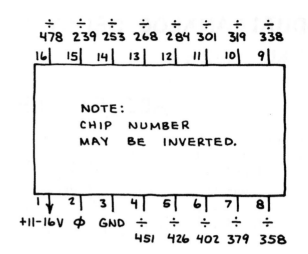

NOTE:
CHIP NUMBER
MAY BE INVERTED.

ADJUSTABLE OCTAVE SYNTHESIZER

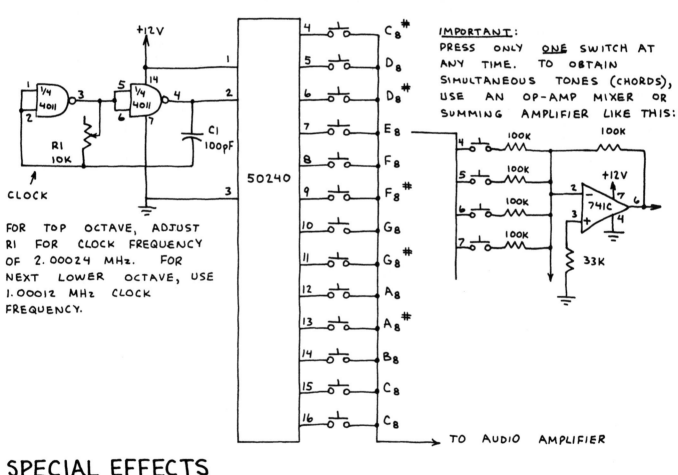

FOR TOP OCTAVE, ADJUST R1 FOR CLOCK FREQUENCY OF 2.00024 MHz. FOR NEXT LOWER OCTAVE, USE 1.00012 MHz CLOCK FREQUENCY.

IMPORTANT:
PRESS ONLY ONE SWITCH AT ANY TIME. TO OBTAIN SIMULTANEOUS TONES (CHORDS), USE AN OP-AMP MIXER OR SUMMING AMPLIFIER LIKE THIS:

TO AUDIO AMPLIFIER

SPECIAL EFFECTS

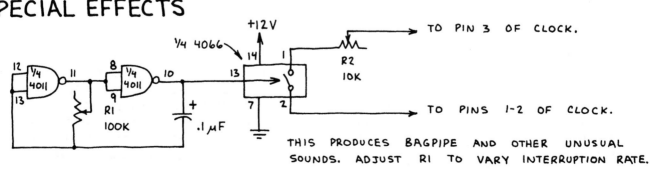

THIS PRODUCES BAGPIPE AND OTHER UNUSUAL SOUNDS. ADJUST R1 TO VARY INTERRUPTION RATE.

146

OPTOCOUPLERS
TIL 111 – PHOTOTRANSISTOR
TIL 119 – PHOTODARLINGTON

INFRARED LED TURNS ON PHOTOTRANSISTOR WHEN LED IS FORWARD BIASED. USE TO REDUCE ELECTRICAL NOISE AND SHOCK HAZARD. IDEAL FOR ISOLATING AND INTERFACING MICROCOMPUTER BUS LINES.

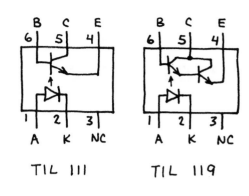

TIL 111   TIL 119

USE TIL 119 WHEN INPUT SIGNAL IS SMALL.

TIL111 / TIL119 TEST CIRCUIT

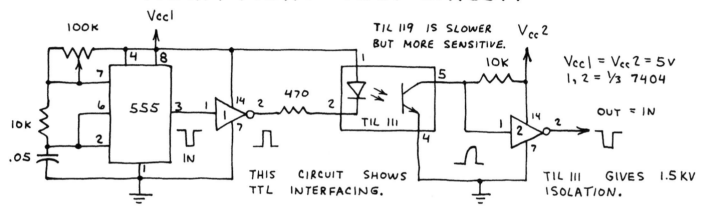

TIL 119 IS SLOWER BUT MORE SENSITIVE.

$V_{cc}1 = V_{cc}2 = 5V$
$1, 2 = \frac{1}{3}\ 7404$

OUT = IN

THIS CIRCUIT SHOWS TTL INTERFACING.

TIL 111 GIVES 1.5 KV ISOLATION.

CALCULATOR / COMPUTER INTERFACING

KEYBOARD INPUT

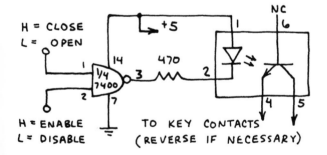

H = CLOSE
L = OPEN

H = ENABLE
L = DISABLE

TO KEY CONTACTS
(REVERSE IF NECESSARY)

IMPORTANT: THESE CIRCUITS MAY VOID YOUR CALCULATOR'S WARRANTY. I HAVE USED BOTH WITH A LOW COST CALCULATOR WITH LED READOUT. SEE POPULAR ELECTRONICS, DEC 1979 (PP. 85-87) FOR DETAILS. ALWAYS FOLLOW MOS HANDLING PROCEDURES WHEN WORKING WITH CALCULATORS! IF NOT, YOU MAY DAMAGE THE UNIT'S PROCESSING CHIP.

CALCULATOR TIMER

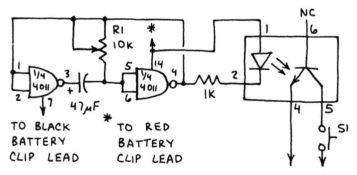

TO BLACK BATTERY CLIP LEAD

TO RED BATTERY CLIP LEAD

TO OPERATE:

1. SET R1 TO GIVE 10 Hz FREQUENCY.

2. ENTER ⊡ 1 ⊞

3. PRESS S1 FOR TIMING PERIOD.

4. READ TIME TO TENTH SECOND FROM DISPLAY.

TO ⊟ KEY CONTACTS

NOTE: THIS SHOWS CMOS INTERFACE.

147

OPTOCOUPLERS
MOC3010 - SCR
SCS11C3 -TRIAC

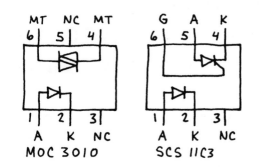

MOC3010   SCS11C3

INFRARED LED SWITCHES TRIAC (MOC3010) OR SCR (SCS11C3). MOC3010 WILL SWITCH 120 VOLTS AC AT 100 mA. SCS11C3 WILL SWITCH 200 VOLTS DC AT 300 mA.

CALCULATOR OUTPUT PORTS

SCR (DC) PORT

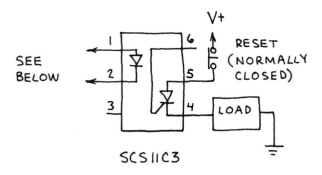

SCS11C3

TRIAC (AC) PORT

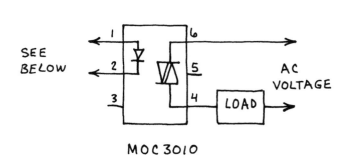

MOC3010

CONNECT PINS 1 AND 2 TO DECIMAL POINT OF LOWEST ORDER READOUT DIGIT. BE SURE TO OBSERVE POLARITY. USE ONLY WITH CALCULATOR HAVING LED READOUT. TYPICAL OPERATION: KEY IN NUMBER WHICH PLACES DECIMAL ANYWHERE BUT FINAL DIGIT. THEN PRESS [−] [1] [·] [0]. NUMBER IN DISPLAY WILL BE DECREMENTED EACH TIME [=] IS PRESSED. WHEN COUNT REACHES 0, DECIMAL MOVES TO LAST DIGIT AND ACTUATES OUTPUT PORT. FOR MORE INFORMATION SEE <u>POPULAR ELECTRONICS</u>, DEC. 1979 (PP. 86-87). SOME CALCULATORS WILL REQUIRE DIFFERENT KEYSTROKE SEQUENCE. <u>IMPORTANT</u>: THESE CIRCUITS MAY VOID THE WARRANTY OF YOUR CALCULATOR OR COMPUTER. FOLLOW MOS HANDLING PROCEDURES TO AVOID DAMAGING CALCULATOR OR COMPUTER. COMPUTER PORTS DESIGNED TO INTERFACE WITH TTL OR LS BUS LINES.

THE LOAD FOR ALL THESE CIRCUITS MAY BE LAMP, MOTOR OR OTHER DEVICE WHICH DOES <u>NOT</u> EXCEED RATING OF OPTOCOUPLER.

COMPUTER OUTPUT PORTS

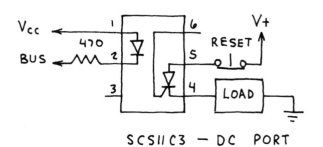

SCS11C3 — DC PORT

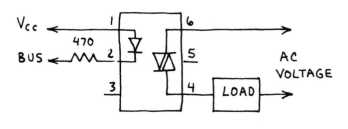

MOC3010 — AC PORT

OPTOCOUPLER
MOC5010 LINEAR AMPLIFIER

CONVERTS CURRENT FLOW THROUGH
LED INTO OUTPUT VOLTAGE.
IDEAL FOR TELEPHONE LINE
COUPLING AND VARIOUS AUDIO
APPLICATIONS.

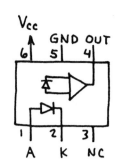

ISOLATED ANALOG DATA LINK

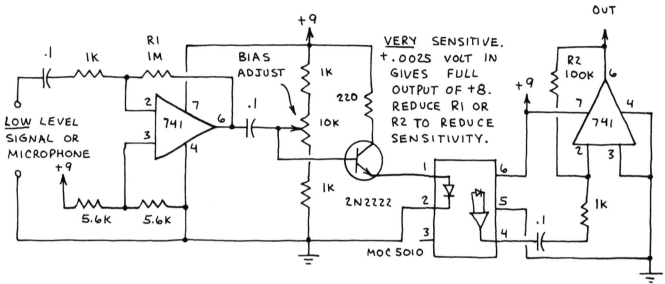

LOW LEVEL
SIGNAL OR
MICROPHONE

BIAS ADJUST

VERY SENSITIVE.
+.0025 VOLT IN
GIVES FULL
OUTPUT OF +8.
REDUCE R1 OR
R2 TO REDUCE
SENSITIVITY.

2N2222

MOC 5010

SCR DRIVER   ## TTL INTERFACING

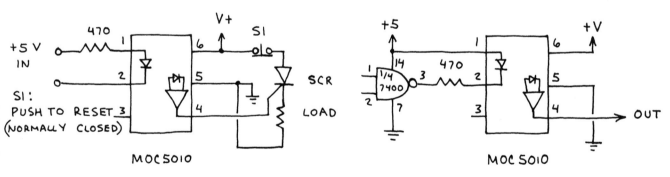

+5 V
IN

S1:
PUSH TO RESET
(NORMALLY CLOSED)

SCR
LOAD

MOC5010   MOC5010

AC SIGNAL ISOLATOR

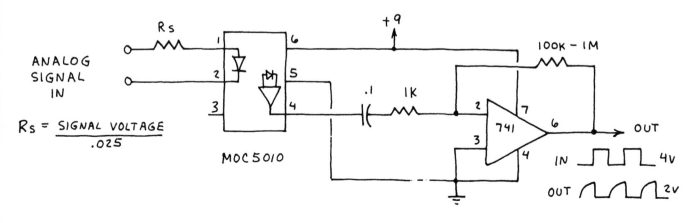

ANALOG
SIGNAL
IN

$$R_S = \frac{SIGNAL\ VOLTAGE}{.025}$$

MOC5010

IN ___ 4V

OUT ___ 2V

149

NOTES

NOTES

INTEGRATED CIRCUIT INDEX

NOTE: TTL and LS
chips are generally
interchangeable.
LS chips consume
less power than TTL
equivalents. Use LS
chips for battery-
powered circuits.

NOTE: The Linear
index includes some
CMOS/MOS chips.

INDEX OF CIRCUIT APPLICATIONS

154

Printed and bound by CPI Group (UK) Ltd, Croydon, CR0 4YY

08/05/2025

01864912-0002